D0215419

THE ALIVENESS OF PLANTS: THE DARWINS AT THE DAWN OF PLANT SCIENCE

THE ALIVENESS OF PLANTS: THE DARWINS AT THE DAWN OF PLANT SCIENCE

BY

Peter Ayres

LONDON
PICKERING & CHATTO
2008

Published by Pickering & Chatto (Publishers) Limited
21 Bloomsbury Way, London WC1A 2TH

2252 Ridge Road, Brookfield, Vermont 05036-9704, USA

www.pickeringchatto.com

BRITISH LIBRARY CATALOGUING IN PUBLICATION DATA

Ayres, P. G. (Peter G.)
The aliveness of plants: the Darwins at the dawn of plant science
1. Darwin, Charles, 1809–1882 2. Darwin, Francis, Sir, 1848–1925 3. Darwin, Erasmus, 1731–1802 4. Botanists – Great Britain – Biography 5. Botany – Great Britain – History – 18th century 6. Botany – Great Britain – History – 19th century
I. Title
580.9'22'41

ISBN–13: 9781851969708

This publication is printed on acid-free paper that conforms to the American National Standard for the Permanence of Paper for Printed Library Materials.

Typeset by Pickering & Chatto (Publishers) Limited
Printed in the United Kingdom at the University Press, Cambridge

CONTENTS

'It might be called, "On the aliveness of plants" which I think sounds more attractive than the life of plants'.

Francis Darwin replies to William Rothenstein's invitation to give a talk to a village society, Monday 23 March 1914 (Rothenstein Letters, Houghton Library, Harvard University).

ACKNOWLEDGEMENTS

Francis was the first Darwin to whom I was introduced. We met when I writing a biography of Harry Marshall Ward, the distinguished plant pathologist. The two men were of similar age and met when young, while they were learning their trade – botany. Their friendship was sustained throughout their lives and for the last years of Harry's life they were colleagues in Cambridge. I must therefore thank Harry for kindling my interest in the Darwin family. My interest in Francis owes much also to my own friend and colleague at Lancaster University, Terry Mansfield, an eminent stomatal physiologist. My thanks go to Terry for reminding me often that Francis was a pioneer of stomatal physiology and that he and his father, Charles, were pioneers in the study of plant movements. It was clear to me that in order to understand Francis's achievements I had to know how much he had been inspired by his illustrious father. And for comparable reasons, Charles led me in turn to his grandfather, Erasmus. Where their botany was concerned, the Darwins could not be separated.

Desmond King-Hele's and Jenny Uglow's writings were wonderful guides to Erasmus, but I think my enthusiasm for Darwins might have faded if it had not been for the personal encouragement of Janet Browne, that great authority on Charles. She welcomed me, an outsider, to the world of Darwin students.

Among other Darwin experts who kindly pointed me in the right directions in Cambridge, where most of the archives are to be found, were John Parker (Plant Sciences) and staff of the Darwin Correspondence Project (University Library) particularly Gina Murrell, Alison Pearn and Andrew Sclater. Also in Cambridge, and due my thanks for their researches on my behalf, are Christine Alexander (Plant Sciences), Jacqueline Cox (University Archives), Colin Higgins (Christ's College) and Anne Thomson (Newnham College). Torri Reeve (English Heritage), Stephanie Jenkins (local historian), and Stephen Foster and Vic Anderson (Southampton University): each tracked down for me information about the personal lives of the Darwins. When I was learning about Ingen-Housz, Sachs and Henslow, botanists whose lives impinged significantly on those of the Darwins, Norman and Elaine Beale, Howard Gest, Wolfram Hartung and Max Walters gave me invaluable advice and material help. So, too, did Peter Lea about Lawes, Gilbert and the early days of Rothamsted Experimental Station.

At Lancaster University, Alan Shirras (Biological Sciences), Sue Clarke and Jacqueline Whiteside (Library) helped me access the Rothenstein Collection of letters in the archives at the Houghton Library, Harvard University. Closer to home, staff in the libraries of Oxford University (old Bodleian, Corpus Christi College, Somerville College, and especially Plant Sciences and the Radcliffe Science Library) have been unfailingly friendly and helpful.

Finally I thank my wife, Mary, for her encouragement. Her interest and practical help during numerous fact-finding expeditions gave me the energy to complete the project.

LIST OF TABLES

LIST OF FIGURES

Unless otherwise stated, all figures are either out of copyright or are original works of the author.

Frontispiece: Four generations of Darwins. All were tall, solidly built men. Erasmus (top left) became so large that a semicircle had to be cut to accommodate him at the dining table. Robert Waring (top right) weighed twenty-four stone near the end of his life. Charles (bottom left) and Francis (bottom right) were equally broad shouldered but not overweight. Charles developed a pronounced stoop as he grew older. Reproduced by permission of the Wellcome Photolibrary (Erasmus, Robert and Charles) and the Botany School, Cambridge (Francis).

1 GREEN THREADS ACROSS THE AGES: A BRIEF PERSPECTIVE ON THE DARWINS' BOTANY

Plants have determined man's history and they will determine his future. Whether they were suspended in the primeval soup of the oceans, or more firmly anchored after their emergence onto land, plants modified the earth's early atmosphere, fixing carbon dioxide (CO_2) and releasing oxygen (O_2), making possible the evolution of primitive animal life and, much later, of man himself.[1] Plants are at the base of every food chain, supplying us with both building and clothing materials, with wood for cooking and heating, and with fuels to generate power. Such green threads connect our most fundamental activities yet, despite this, we have until very recently taken plant growth and health for granted. Many aspects of plant physiology remain a mystery to us, particularly where they concern plants under the variable conditions of the field – away from the controlled conditions of the laboratory. Still in its infancy is the study of how the physiology of an individual plant is modified when it functions as a member of a community in a natural environment. One concern of the greatest current interest is the extent to which plants, with their capacity to fix CO_2, might ameliorate changes in the composition of the earth's atmosphere that threaten our safe and familiar world.

Given the importance and urgency of the latter, it is salutary to review the progress we have made in our scientific understanding of plant growth and to reflect that our understanding began less than three centuries ago. And this is where the Darwin family comes in.

A strong green thread ran through that family, connecting different generations across two hundred years. Erasmus Darwin (1731–1802), his grandson, Charles (1809–82), and his great grandson, Francis (1848–1925), were all botanists of distinction, each fascinated by the movements of plants as they seek light, nutrients and water.

Mention of the name Darwin has for too long conjured an image of just one of them, Charles, his theory of evolution, fossils, struggles for survival in the animal world, and man's origins with the apes. As this book will reveal, the image is misleading where Charles in concerned. It is moreover unfair both to Erasmus and, particularly, to the previously forgotten Francis, each of whom deserves to be better remembered.

The popular image of Charles is misleading simply because he studied plants as much as animals, finding their study satisfying and rewarding. He wrote that 'It has always pleased me to exalt plants in the scale of organised beings' and, in a letter to J. D. Hooker dated 3 June 1857, that he found 'any proposition more readily tested in botanical works ... than zoological.[2]

It is unfair because Erasmus led 'a life of unequalled achievement',[3] not least when he was popularizing botany and writing about plant function with extraordinary prescience. It is unfair because Francis continued and with great vision enlarged upon Charles's botanical works. The Darwins' botany has borne fruit in many areas; its outcomes are easily recognizable in many of today's horticultural and agricultural practices (see Chapter 11). But what is most remarkable is that the culmination of the Darwins' fascination with movements, and its extension by Francis's young colleague in Cambridge, Frederick Frost Blackman, has provided the very foundations upon which our modern studies of interactions between plants and the atmosphere are based. Thus, Francis was a pioneer in the study of the movements of stomata – pores in the otherwise impermeable surface of the leaf whose daily opening and closing regulates entry of CO_2 into the leaf, and the escape of water from it (Figure 1.1) – while Blackman complemented Francis's work by defining the way in which the levels of key environmental factors, CO_2, temperature, and irradiance limit the rate at which carbon is fixed.

To understand how water was lost by transpiration – or perspiration as Erasmus called it – and how plants obtained most of their dry weight from the air (not from the soil or, indeed, by transforming water) proved fundamental to understanding their growth, that most basic task facing those early botanists

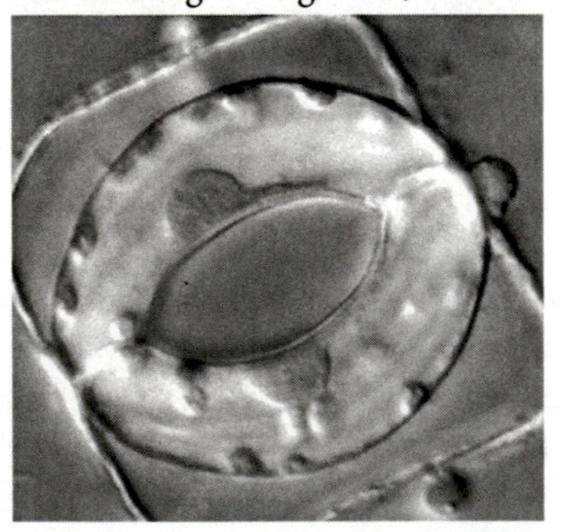

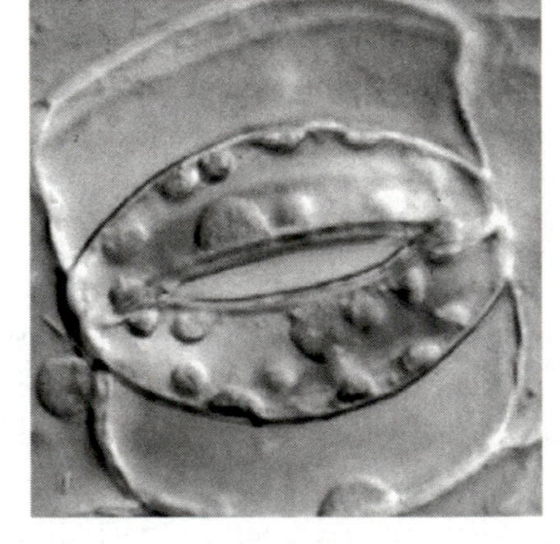

Figure 1.1: In epidermis detached from a leaf of *Commelina communis*, an open (left) and almost closed stomatal pore (right) are seen from above. Each pore is surrounded by two kidney-shaped 'guard cells' whose swelling and shrinking determine the pore area. Reproduced by permission of Dr Mark Fricker, Plant Sciences University of Oxford

interested in more than merely naming and describing plants. Later chapters will describe the progress they made and the connections they revealed between water and movements. But why should those basic subjects, transpiration and photosynthesis, be so important for today's botanists (or plant scientists as they prefer to be known)? At the heart of the answer lies the subject of energy and its transformation. Pigments in plants, such as the green chlorophylls, trap solar energy in the first stage of a process that has become known as known as photosynthesis. In the second stage, which is independent of light, that energy is used to drive a series of chemical reactions during which CO_2 is used, with other chemicals, to make sugars and energy-rich compounds, respectively, the building blocks and fuel needed for growth.

Photosynthesis occurs in specialized bodies, chloroplasts, located deep within the cells of leaves and green stems. In order to get from the atmosphere to those cells, CO_2 has to diffuse through stomatal pores in the epidermis or surface layer of cells of the plant. The composition of our atmosphere has been continually changing over millions of years. Today its concentration of CO_2 is around 382 parts per million (ppm = 0.0382 per cent) and rising fast, perhaps by as much as 2 ppm per year.[4] In a recent but pre-industrial age it was 273 ppm. Our way of life is threatened because CO_2 in the atmosphere stops some of the solar energy that strikes the earth's surface from being immediately re-radiated back into space; the higher the concentration of CO_2 in the atmosphere the more energy that is retained. The energy is trapped in the form of heat under the thickening blanket. Higher CO_2 levels are not in themselves a direct threat to life but they contribute to global warming. Among the forecast consequences of warming are large but unpredictable changes in climatic factors, such as amounts of rainfall, probably leading to hugely increased areas of aridity that are totally unsuitable for plant growth which, of course, would in turn affect climate.

And here is a conundrum: at higher CO_2 concentrations more carbon dioxide diffuses into the leaf, and the rate of photosynthesis is higher, than at lower more 'normal' concentrations. Furthermore, at high CO_2, less water is lost per unit of carbon captured – in other words 'water use efficiency' is increased. So, in theory, plants *can* cope with anthropogenic rises in CO_2 levels, and even ameliorate those increases, but the question is, *will* they do so in practice? We cannot yet answer this question with any certainty because we are still learning about the limits of plants' capacity to respond to ever increasing CO_2 concentrations. It is clear that while photosynthesis in most plants increases as CO_2 levels rise to about 600 ppm – if water is freely available and the temperature is suitable – photosynthesis shows less and less response as concentrations rise above 600 ppm. Even below 600 ppm there are variations between plant types, with many annual crop plants responding to higher CO_2 levels less than woody crops, such as cotton, or many tree species, for reasons that are not yet entirely clear.[5]

The Darwins' individual contributions to the evolution of botany (plant science) are laid out in the following chapters, as are the connections between the three men. Each man has his own intrinsic interest but, seen in a wider context, the differences between their working lives illustrate contemporary sociological changes that were affecting the study of plants, such as the battle between 'country house' and laboratory science, the professionalization of science, and the rise of women in the laboratory. The story of the Darwin family is the story of the evolution of plant science.

The Darwins were not *the* founders of plant science but individually and collectively they made an unique contribution to its early development. Along with a handful of others, they played major parts in the growth and sustenance of its core area – physiology – which is concerned with the functioning of organs or systems in relation to their structure. Quantitative, that is measurement-based, experimental plant physiology began with the Reverend Stephen Hales in 1727. Its forward march stuttered through the next hundred years, but then accelerated. The scientific story originates therefore with Hales and ends with the death of Francis Darwin in 1925, or, to be more exact, a few years *after* his death, for, in order to appreciate the Darwins' scientific legacy, it is necessary to follow forwards in time the particular subjects on which they worked. What is indisputable is that by the period between the two world wars botany had become an independent and rigorous, laboratory-based science.

Before exploring the botany of the Darwins, however, it is necessary know more about the family and its fortunes, for these strongly influenced the ways in which Erasmus, Charles and Francis approached the study of plants. Among those who influenced the Darwins, or were influenced by them, one of the most significant was the German botanist, Julius von Sachs. Like him, I trust that, 'I may sometimes have overrated the merits of distinguished men, but have never knowingly underestimated them'.[6]

2 THE FORTUNES OF THE DARWINS

In the eighteenth and nineteenth centuries, a man of outstanding talent might sometimes achieve success even if his origins were humble, but he stood a much better chance of success if from birth he was assured of financial security and a good social position. The Darwins were blessed with both security and position. They had no need of patronage. Their exceptional talents were free to blossom in whatever sphere of endeavour they chose, their comfortable well-regulated lives being disrupted only by ill health and, occasionally, by an untimely death.

Plants were a hobby for Erasmus Darwin, the busy general practitioner in Lichfield, a small town in the English Midlands some fifteen miles from the city of Birmingham. They formed an essential part of the studies of Charles, the gentlemen natural scientist, secluded in his large house in the quiet countryside of Kent, and they were the basis of a profession for Francis, the laboratory scientist who did his best work in a university environment. The transition during those two centuries from Erasmus the theorizer to Francis the experimenter mirrored changes happening in the wider world of botany, and happening too in most of the nascent sciences. The significance of those changes is explored in later chapters, but first we need some perspective. This chapter explores the personal lives of Erasmus, Charles and Francis because, as for all men, the twists and turns of their careers, their triumphs and failures, were rooted in the background of their family, in the human cycle of birth, marriage and death.

*****

The Darwins' capacity to regulate and control their lives stemmed from the family's modest wealth which was itself derived from ownership of estates in Lincolnshire and Nottinghamshire since at least the seventeenth century. Indeed, ownership may have stretched much further back for the family was granted heraldic arms in the sixteenth century.[1] This pattern of squiredom was disturbed early in the eighteenth century by one Robert Darwin who tried his hand, rather unsuccessfully, as a lawyer in London before retiring to Elston, just north east of Nottingham, to adopt a life of relative indolence. This particular Robert – there were others of the same name to follow – is noteworthy only for his lineage: his seventh and youngest son was Erasmus, born in 1731 and destined to become a distinguished doctor, poet, and 'philosopher' (the nearest modern equivalent is 'thinker').[2]

Erasmus was in the words of his biographer, Desmond King-Hele, 'a man of unparalleled achievement'.[3] It was the multi-talented Erasmus who gave the Darwin family its unique identity, for it was he who was the first of a long line of intellectual, albeit practical, men. The Darwins who followed Erasmus were 'natural historians' – only later known as 'scientists' – and mathematicians. Erasmus was the first member of the family to be elected a Fellow of the Royal Society but, from that occasion in 1761, there followed a continuous succession of Darwins, including Charles and Francis, who were selected by their peers to receive the same accolade, the highest in British science. (Erasmus was elected on the strength of two papers published in the Society's *Philosophical Transactions*. The subjects of the papers illustrate the breadth and diversity of his interests, the first describing an experiment to demonstrate the effects of solar radiation on the expansion and density of vapours,[4] the second outlining a procedure to avoid congestion of the lungs, a procedure which owed everything to a simple physical interpretation of lung function.)

As a child, Erasmus joined in boisterous games with his siblings, but he seems from an early age to have preferred poetry and experiments to exercise and country sports. These signposts to his later life became even clearer when, at the age of ten, he was sent to school in Chesterfield. With a fellow pupil, Lord George Cavendish, he almost blew up the school while experimenting with gunpowder while, at the same age, sending to his sisters at home letters that frequently broke from prose into poetry.

All who knew Erasmus in later life remarked on his huge presence. Both tall and, by mid-life, corpulent and clumsy, he was a man of exceptional appetites, for food, wine – at least in his youth – and women. An anonymous friend from his student days in Edinburgh remarked,

> in his youth, Dr Darwin was fond of sacrificing to both Bacchus and Venus; but he soon discovered that he could not continue his devotions without destroying his health and constitution. He therefore resolved to relinquish Bacchus, but his affection for Venus was retained to the last period of life.[5]

Erasmus was married twice, to Mary (known as Polly) and Elizabeth, and fathered several children outside marriage, including, by his long-time mistress, Mary Parker, two girls whom he openly acknowledged, educated and supported into their adulthood (he was, in advance of his times, a strong advocate of education for females).

Above all, Erasmus had an insatiable appetite for intellectual stimulation, for curiosities and for work. From boyhood, he was fascinated by mechanics and chemistry, making an alarm for his fob watch and dabbling in experiments involving electricity. In later life, he designed a windmill for grinding pigments for his good friend, the potter, Josiah Wedgewood I. He modified the design of his own horse-drawn carriage and had ideas for a steam carriage. He even designed a 'speaking machine' to be built from wood, with bellows that pumped air across a silk ribbon to reproduce the effects of the vocal chords. Typically, he never finished the last project, nor did he follow to completion many of his ideas, for he enjoyed speculation, plans and invention as much as their practical outcome. There was far too little time left over from his medical practice to allow him to follow through every idea thrown up by his fertile imagination and he was not motivated to exploit his inventions for profit.

Erasmus was the first of several Darwins to study medicine. As was not uncommon at the time, he viewed the human body as a machine, believing it obeyed the same Newtonian rules as did his watch and his carriage. He was a popular and successful general practitioner in Lichfield, and also in Derby where he moved much later in his life. Erasmus did not grow rich from medicine, particularly since his hobbies and his endless travels to meet and talk with men of like minds and energies were a significant drain on his finances but, with his inherited wealth, he was more than comfortable. It seems he struck a nice balance between generosity and hard-headedness, having a reputation for giving his professional time free to the poorest men and women who came to him, while overcharging his richer patients.

For all his generosity, and for all the love and admiration that was poured upon Erasmus by his patients and fellow intellectuals, there was a darker side to his character. He could be exceedingly sarcastic, often stubborn and overbearing. Notably, his relationship with his third son, Robert Waring, was at times difficult. When in 1778 his first son, Charles, who was training to be a doctor at Edinburgh University, cut his finger during a dissection and died of the resulting septicaemia, Erasmus decided that Robert, then barely twelve years old, should take over the medical mantle. The problem was that although Robert had adored his eldest brother and shared his love of natural history, he hated medicine. He hated it then, he hated it when at age fifteen he was despatched to London and then Edinburgh to begin his studies, and he hated it to the end of his days, always feeling revulsion at the sight of wounds and blood – a horror which he transmitted to his own son, Charles.

Erasmus was not to be deflected. Having virtually forced Robert into medicine he then proceeded to interfere quite shamelessly in the promotion of his son's medical career, chivvying the professors in Edinburgh to favour him, finding openings for him in Leyden and Paris and, finally, persuading the Royal Society to elect Robert to a Fellowship in 1788. Erasmus must have been bitterly disappointed when Robert settled for the potentially quiet life of a general practitioner in Shrewsbury, for it was not the glittering career path that Erasmus had in mind for him. Nevertheless, Erasmus gave him £20 to set him on his way – as too did his Uncle Robert – and the young man soon established what proved to be a highly lucrative practice, possibly the richest outside London.[6] One factor in his favour was that, like his father, he was exceptionally popular with the opposite sex. He soon had many ladies amongst his patients. He quickly gained their confidence and they poured out their hearts to him.[7] Another factor in his favour was that at this critical time he received two small inheritances, one from his mother and one from an aunt. If he did not gain satisfaction from medicine, he certainly found it elsewhere – in making money. It was he who transformed the Darwin family's finances from what could have been regarded as 'sound' to what could be described as 'extremely comfortable'.

As soon as Robert was able, he began to invest in property, first domestic and then commercial. From the high rents he charged to tenants, he was able to move into more lucrative ventures. He saw the opportunities offered by investment in the new roads and canals that were so essential to the burgeoning industries of the midlands and north, and he reaped rich rewards. Both his security and the capital at his disposal increased enormously when he married Susanna, eldest daughter of the potter Josiah Wedgwood I, for she brought to the marriage a fifth share of the Etruria works and, upon the death of her father in 1795, no less than £25,000 (the equivalent of over one million pounds today).[8] In the late eighteenth century banks were still small independent concerns, in their activities barely distinguishable from money-lenders. Many local entrepreneurs, as well as some hard-up gentry, were happy to turn to Robert rather than to a bank for a loan. He moved his money rapidly from one enterprise to another, with an unerring eye for profit. He helped the Wedgwoods through a critical period by buying shares in their pottery and in 1802 he lent money to Josiah II that enabled him to buy Maer Hall. By the time he died in 1848, Robert weighed twenty-four stone and was worth £223,759. It was said that he owned more than half of the sizeable market town of Shrewsbury. It was because of Robert's genius for money-making, and his marriage, that Charles and his own sons never needed to work for money.

Although Charles lacked his father's aggressive risk-taking attitude to investment, he was, after his own fashion, wise and steady. He moved money in and, just as judiciously, out of the stocks of the fast-expanding railway companies but he favoured the safety of gilt-edged securities and property. Among those properties, he received regular income from the farm in Lincolnshire of which the family still retained ownership. Again in character, Charles kept the most detailed records of not just his investments and the income they generated but also of day-to-day household expenditure. If specialist advice was needed then, in later life, it could be supplied free of charge by his eldest son, William, who became a banker. Francis recalled how, 'he used often to say that what he was really proud of was the money he had saved'. The income he derived from his writings gave him special pleasure.[9] There was justifiable reason for pride; William estimated that, shortly before Charles's death in 1882, his father's capital amounted to £282,000 (the equivalent of around £13 million now). His wife, Emma, whom he married on 29 January 1839, had her own small fortune. She was one of eight children of Josiah Wedgwood II (making her a cousin to Charles), the son who had taken Josiah I's business from strength to strength.

Where passing wealth onto his children was concerned, Charles, unlike his grandfather Erasmus, was generous and even-handed, with the exception that he discriminated between the boys and the girls. For the last few years of his life he gave an annual allowance to each son; in 1881 this amounted to £474 (roughly equivalent to £20,000). There were also handsome settlements upon marriage; Francis's amounted to £5,000. After Charles's death, the boys benefited not only from their father's estate but also from the estate of the aptly named Anthony Rich. This wealthy, childless solicitor had been one of Charles's greatest admirers. It was his strongest wish to leave his personal fortune to Charles, but Charles had been equally firm in his polite refusal to accept such a bequest. Thanks to the mediation of T. H. Huxley who smoothed the way for Charles, as in so many difficult issues, a happy compromise was reached; Rich's will was written in favour of Charles's five sons. Thus, when Rich died in 1891, nine years after Charles, each of the Darwin boys inherited several thousand pounds.

*****

Where health was concerned, the Darwins were less fortunate, illness and death profoundly affecting the shape of the their lives. Erasmus appears to have enjoyed robust health, in spite of his corpulence. His energy and spirit gave him the resilience to overcome the loss of his first wife, Polly, to cirrhosis of the liver in 1770 and of his beloved first son, Charles, to septicaemia in 1778.[10] He was not a man to dwell on misfortune. By contrast, ill health – either actual or anticipated – was a constant preoccupation for Charles in his middle and later years and, following his example, also for his children.

The wealth that Charles had inherited, and which he had nurtured so astutely with meticulously planned investments, enabled him to protect himself and to cocoon his family in the secure, comfortable surroundings of the home, Down House, that he and Emma had in 1842 found on the edge of the village of Downe, deep in the Kent countryside but less than twenty miles from central London. At this stage in his life, only six years after returning from the *HMS Beagle*'s five-year long voyage around the southern oceans, and aged just thirty-three, he had already lost his youthful spirit of adventure. From early middle age onwards, and suffering for the rest of his life digestive problems that were never satisfactorily diagnosed – and which still remain mysterious – he rarely left Downe, travelling only when there was a compelling reason. Even these few journeys exacted a price, for they were often followed by bouts of sickness and exhaustion. Except for the company of a very few old and trusted friends, such as Joseph Hooker and his family, he could find complete relaxation only with members of his own family. This self-imposed seclusion had a positive and important outcome, however, because through it he was able to clear for himself the uncluttered time and space that he needed for his research and writing.[11]

In ill health, Charles had also found a justification for letting others face the public when the time came to present his theory of evolution by natural selection. As Naturalist on the *Beagle*, studying the flora and fauna of South America and the Galapagos islands, he had begun to assemble the evidence and arguments that he finally presented to the world in *On the Origin of Species*, but publication resulted only many years later, in 1859.[12] His reluctance to publish stemmed, partly, from a fear that the scientific community would not accept his biological and geological evidence and, partly, from a fear that the wider community of intellectuals would be outraged because his ideas undermined Christian teaching. Although natural evolution remains unacceptable to a few men and women in the twenty-first century for religious reasons, and scientists from ecologists to molecular biologists still subject the proposed mechanisms to intense scrutiny, Charles's arguments were – once the initial furore had died down – more readily accepted into the orthodoxy of science than he had dared to hope. No serious scientific challenge was mounted until the first studies of genetic mutations were made in the early years of the twentieth century. Contemporary scientific opinion was won over more easily than Charles expected because the worlds of biology and geology had for decades, or even centuries before, slowly been edging towards accepting the idea that changes had occurred over a time-scale much longer than that acknowledged by the established churches. His grandfather, Erasmus, may have anticipated the theory of organic evolution, as will be seen in the next chapter, but others too had recognized small pieces of the jigsaw. The evolution of plants, animals, rocks and soils was therefore not a totally new concept. Charles's unique achievement was to present both the detailed evidence and the plausible mechanisms that were needed to construct a robust integrated theory.

Charles's fears were similarly unfounded where *On the Origin of Species*'s reception by the wider literate public was concerned, for their mood was generally sympathetic. The simplified picture of 'nature red in tooth and claw', famously painted by Alfred Lord Tennyson in his poem *In Memorium A. H. H.* (1850),[13] was already well embedded in the minds of educated men and women and it was readily applied by them to Darwin's theories. The public had been well primed. Tennyson's vision however was essentially a zoological one. It provided a graphic image for what in the popular imagination has all too often over succeeding generations been regarded as the only thrust of Charles's arguments for natural selection, namely, fitness is associated solely with predator/prey relationships. Such dramatic contests did form an essential part of his thesis and were readily understood by the majority of his readers but, in addition to the zoological arguments relating to fitness, Charles drew his evidence much more widely, detailing in *On the Origin of Species* several examples of subtler struggles involving plants. The force of these illustrations has, however, frequently been overlooked and they have remained the province of specialists.

Charles's more general arguments likewise often related to plants. Thus, in reply to his critics' argument that organic beings had been created beautiful for the delight of man, Charles countered:

> If beautiful objects had been created solely for man's gratification, it ought to be shown that before man appeared, there was less beauty on the face of the earth than since he came on stage ... Flowers rank among the most beautiful productions of nature; but they have been rendered conspicuous in contrast with green leaves, and in consequence at the same time beautiful, so that they may be easily observed by insects ... if insects had not developed on the face of the earth, our plants would not have been decked with beautiful flowers, but would have produced only such poor flowers as we see on our fir, oak, nut and ash trees, on grasses, spinach, docks, and nettles, which are all fertilized through the agency of the wind.[14]

Through much of Charles's working life his thoughts were occupied with plants rather than animals. From about 1855, nineteen years after he returned from the voyage of the *Beagle* and four years before publication of the first edition of *On the Origin of Species*, until his death in 1881, he occupied himself with a continuous stream of botanical projects.[15] Finding inspiration in one group after another, he published several articles and five major books based almost entirely on his own original researches, *The Various Contrivances by which Orchids Are Fertilised by Insects and the Good Effects of Intercrossing* (1862), *The Movements and Habits of Climbing Plant* (1865), *Insectivorous Plants* (1875), *Effects of Cross and Self-fertilization in the Vegetable Kingdom* (1876) and *The Power of Movement in Plants* (1880). Each book has its own lasting, intrinsic value but their contemporary significance was that they provided Charles with material, for example concerning the evolution of the orchid flower (see Chapter 5), that he could use to bolster and extend arguments he had made in the first edition of *On the Origin of Species*. The final edition of *The Origin* ('*On*' had been dropped from the title) was thus significantly longer and more detailed than the first.

Each article and book was packed with original observations. Charles presented a wealth of new information that has stood the test of time simply because his observations were both detailed and accurate. The value of Charles Darwin's botanical works has rarely been appreciated. If he had produced no other works than these, posterity would have judged him a scientist of the first rank. However, the greatest of these botanical works, *The Power of Movement in Plants*, might never have been started if it had not been for a tragic event, the death of Francis's young wife, Amy. Allied to Charles's grief there was a sense of shock and a feeling of impotence, for while he could control most aspects of his life thanks to his wealth and his connections, he was painfully reminded that, as when his own daughter, Annie, died, or when he himself had suffered one of his recurrent bouts of sickness, he could not use his wealth to guarantee good health. Even the best and most highly paid doctors in the land were powerless to help. Charles could not protect those dearest to him.

It is worth dwelling on those events of the summer of 1876 because they had such profound repercussions for the last years of Charles's career and virtually all of Francis's. Amy had been just twenty-six years old when she died. Baby Bernard was her first child and, although he had been born healthy, she had developed puerperal fever. Her temperature rocketed as the infection took hold and at its height she suffered hallucinations and terrible convulsions. Francis would not leave her bedside but, as a qualified doctor, he saw all too clearly that nothing could save her. And so it proved; only two years after Francis and Amy had married, he was a widower.[16]

The death of Amy was a devastating blow to Francis, from which he never fully recovered. For the remainder of his life he was prone to fits of depression. He was constantly plagued with worries about Bernard's health. Two years later, when working in Würtzburg, he wrote to Amy's mother, 'My old German master is always giving me sentences to translate which makes my blood run cold, such as "When I came to my house, I found my son very ill"'. Another time Francis wrote, 'I think I should be helpless with fear if he [Bernard] were ill'. He was terrified whenever one of the women in the family was about to give birth, kept his distance from the event, and only recovered his normal composure after she was safely delivered.[17]

Amy's death was a devastating blow, too, to Charles and Emma. For the sixty-seven-year-old Charles, the passing of Amy painfully revived all too fresh memories of the lingering death of his own dear daughter, Annie. She was the daughter who would climb upon his knee and stroke his beard, who would laughingly pirouette around him in the garden, and who would lovingly slip her small hand into his. He had loved her for her neatness and her skills with the needle and the pen, but, above all, he had loved her for the spontaneous affection she showed him. She had died a quarter of a century earlier, in 1851, but the memories of those terrible days were still sharp. Annie had contracted scarlet fever in 1849 and never fully regained her strength. In the spring of 1851, Charles had decided on a visit to Dr Gully's 'water clinic' in Malvern hoping it might prove the tonic needed to cure her (Charles had often visited Gully in a quest to cure the problem of his own 'accursed stomach', sometimes with remarkable success, albeit short-lived).[18] With Emma remaining at Downe where she was expecting the birth of their ninth baby, Horace, Charles and Annie set off full of hope. Charles' letters home sadly became ever more frightened and depressed. Annie's condition was getting worse rather than better. She rallied briefly but suddenly relapsed, dying from a 'bilious fever with typhoid character'. Charles would never know whether his genes or his decision to take Annie to Malvern had contributed to her death, but for ever afterwards he was haunted by feelings of guilt.

What was so poignant about Amy's death was that, although Charles had known her only as a young adult, she had in so many ways reminded him of Annie. Amy had the same openness and spontaneity as had his daughter. By birth Amy was half Welsh but 'wholly Welsh in feeling'.[19] The young woman had brought a breath of fresh Celtic air to the staid and predictable lives of the Darwins. Both Charles and Emma loved her for it. The improbable connection between the two families, the one composed of staid thinkers, the other of bold doers, had begun when, in their early teens, Francis and his brother Horace were sent to Clapham Grammar School. There they made friends with Amy's brothers, who were later to become distinguished soldiers in the Ruck family tradition. Soon the Darwin boys were invited to spend part of the school holidays at the Ruck's family home at Pantlludw, near Aberdovey in Merioneth. There they met Amy. Her first favourite was Horace and she records in her diary that she loved talking to him about 'all sorts of rather tremendous subjects, such as religion and immortality'. She wrote, 'I am watching him. His nature seems a richer one than the others – with more passionate feeling, he is more of a poet in nature.'[20] Horace was caught up in the spirit of Pantlludw. He and Amy ran screaming up and down the local sandhills, hand-in-hand. They 'tried experiments with wine and water'.[21] But when she was twenty and he was only seventeen, the age gap proved too great for their relationship to develop into anything more serious.

After Horace went up to Cambridge University, he disappeared from her diaries. Instead, Amy increasingly appreciated the steadier qualities of Francis – or Frank as he was known to friends and family – who had previously featured only briefly in her diaries, and then not always in a complimentary way. For example, on the way home from a long expedition made one day by the young Rucks and Darwins, it was Frank who had unashamedly fallen asleep. What finally attracted Amy to Frank remains largely unexplained, but, as would be expected of Amy, the attraction clearly stemmed from the heart. Shortly before their wedding she wrote to him, 'It is not long now to wait only to me the days drag rather. I can't get away to talk to you – you will remember the lines in Browning to show me, my darling – Frank I always wanted to tell you what a great good I feel it is to have such a love as yours'.[22] On 23 July 1874, she married kind, generous, reliable Frank at All Souls, Langholm Place, London. Romantically, given their shared taste in poetry, it was the same church at which Robert Browning and Elizabeth Barrett had secretly been married in1846.

Amy had first visited Downe long before her marriage. Like Charles, Emma was easily won over by Amy's easy charm. Her lilting Welsh accent reminded

Emma of her own childhood spent at Maer Hall in Staffordshire, close to the Welsh borders. Naturally, Amy had been apprehensive about meeting the great Charles Darwin, but she found her future father-in-law kindness itself. He called her 'Amy', made little jokes and let her 'clean some manuscript' for him. Soon he was treating her like one of his own children, sending her to collect plants and rocks and, when she was back home in Wales, asking her to measure the depth of wormholes for him. Of Amy, Charles wrote, 'I think she was the most gentle & sweet creature I ever knew'. Of her death, the old man wrote, 'It is the most dreadful thing which has ever happened, worse than poor Annie's death, though not so grievous to me. I cannot bear to think of the future.'[23]

If Charles and Emma could find any consolation in such a terrible event it was that they had a healthy new grandson who was living, at least for the moment, at Down House. Emma immediately brought into the house a local girl, seventeen-year-old Harriet Irvine, as a wet-nurse for baby Bernard. She proved so popular that she remained with the Darwins until the end of her working life. Frank left for Wales, where Amy was buried in the churchyard of her father's village of Corris. He stayed for several weeks with the Ruck family sharing his grief with them and recuperating a little but, still emotionally devastated, he needed little persuasion that on his return he and Bernard should move permanently into Down House. Amy's mother, indeed her whole family, would always be welcome at Downe. In the event a deep and genuine friendship developed between Emma and Mary, Amy's mother, based not on grief, for neither woman was given to self-pity, but rather on shared interests and opinions. Each was often a guest at the other's home – even when they were old ladies in the 1890s, Mary continued to visit Downe – and there were occasional meetings on holiday, as in Charles's last summer,1881, when both families were in the Lake District.[24]

Charles and Bernard were devoted to each other. Charles often spoke of the pleasure it was to him to see 'his little face opposite him' at luncheon. He and Bernard used to compare their tastes, as in liking brown sugar better than white. Their private joke was to tell anyone who was willing to listen, 'we always agree, don't we?' In Bernard's autobiography, written over half a century later, his memory of his grandfather was still very clear: 'when I came down to lunch I wanted to have whistling matches with him and formed a low estimate of his powers'.[25] Now in his early seventies, Charles was increasingly tired and, when Bernard was six years old, he died, but those last years of his life had been illuminated by his young grandson.

But to step back a few years, in 1877, and aged sixty-eight, Charles published *The Different Forms of Flowers of Plants of the Same Species* (based on five earlier published pamphlets). At the time, he wrote:

> I do not suppose that I shall publish any more books ... I cannot endure being idle, but heaven knows whether I am capable of any more good work.[26]

The tiredness of old age was sapping his ambition, as too were family problems. He was struggling to find some way of cheering and motivating his grieving son. The solution that slowly emerged was to involve Francis, who was already his secretary, more and more in his current research. A mere two years later, on 17 July 1879, a revived Charles was able to write to Victor Carus, a biologist and the translator of many of his books:

> together with my son Francis, I am preparing a rather large volume on the general movement of Plants and I think we have made out a good many new points and views ... we have been working very hard for some years on the subject.[27]

In 1880 *The Power of Movement in Plants* was published. What had started as an act of fatherly benevolence had ended in a synergistic relationship. Francis had brought to the relationship first-hand knowledge of the most advanced practical techniques of the day, which he had learned as a researcher in animal physiology laboratories in London and, also, on visits in 1878 and 1879 to the plant physiology laboratories in Germany that were currently leading the botanical world. In energy and technical skills he gave to his elderly father at least as much as he received back in terms of detailed botanical knowledge and scientific experience. Both men found new motivation. Their partnership prolonged his father's research life and induced Charles to become a true experimentalist. *The Power of Movement in Plants* was one of the first books to be published in that emerging discipline destined to be known as plant physiology. In their book the Darwins announced fundamental discoveries about tropisms – the bending of roots and shoots in response to light, gravity, and touch – that are still repeated today in textbooks and specialist review articles. They developed methods for investigating tropisms that for decades afterwards were adopted as worldwide standards.

After Charles's death, Francis became his father's biographer and hence the guardian of his reputation, but he also found time to take up a lectureship, later converted to a readership, in botany in Cambridge University. Unlike his great-grandfather and father, Francis was paid directly and specifically for researching and teaching botany: he was, in short, one of a new breed, for he was a professional scientist. Francis continued to research the tropisms of roots and shoots but he used his expertise in a related yet new area. He pursued a topic that had concerned his great-grandfather, Erasmus. Francis studied patterns of opening and closing of stomata, the pores in leaves through which CO_2 enters and leaves the leaf, and through which water is transpired (or 'perspired' in Erasmus's terms). His experiments were highly successful, thanks to his talent for technical innovation.

In Germany, Francis had studied with Julius von Sachs in Germany and there he was exposed to a new philosophy that he later helped nurture back in Britain where it became known as the 'New Botany'. Centred on an empirical approach, and rejecting speculation, it required that observations and measurements should be made, wherever possible under controlled conditions, as in a laboratory, before any advance in knowledge could be accepted. What Sachs advocated for botany was all the more remarkable given the popularity of the *Naturphilosophie* (a romantic philosophy of nature) that still prevailed in Germany beyond the time of Sachs's birth in 1832. Its leading proponents were Schelling and Hegel, but its origins could be traced back to, amongst others, Rousseau. The philosophy was characterized by Schelling's view that Bacon, Newton and Boyle were 'perverting natural sciences', and by Hegel's derision of those scientists who performed experimental work. At its core, *Naturphilosophie* proposed that not until man had learned to understand external nature, through contemplation of his own innermost reality, could he hope to formulate laws of existence. The concept of an unifying 'ideal morphology' was popular and some advocates argued that everything, from the order in the planetary system to the order in the shape of plants, was to be explained in set numerical proportions.[28] There were many forces that combined finally to discredit *Naturphilosophie*, not least among which were the writings of Charles Darwin. In these Man became the result of a descent rather than the pinnacle of an ascent, and humanity was no longer taken as the only living form capable of integral development and, hence, as the sole measure of the respective development of other forms.

*****

Where social position was concerned, there was no hint of aristocratic blood in either the Darwin or the Wedgwood lines and, until Francis and his siblings were knighted in middle age, there were no titles in the family. Nevertheless, in the highly stratified society of mid-nineteenth century Britain, Charles and Emma's children enjoyed a secure and elevated position thanks to the family's wealth and its long history of owning property. The large but unpretentious family home in the village of Downe would alone have been enough to give the family local respect. However, because the Darwins had been landowners for generations, in contrast to the *nouveau-riche* industrialists who were rapidly buying up old country properties, they would be categorized as gentlemen, that is decent, reliable and trustworthy. When Charles and Emma moved into the village in 1842 they were immediately welcomed by similar 'old' families in the neighbourhood who saw the Darwins as equals – if unexceptional equals.

Among another group, the thinkers and writers, the Darwins were far from unexceptional. Thanks to the achievements of their father and great-grandfather, Francis and the other children of Charles and Emma were from an early age used to mixing with the most famous men and women and the greatest intellects of the day. The pattern had been established by great-grandfather Erasmus. For him, as for succeeding Darwins, one route to familiarity with the leading scientists was through their membership of an exclusive club, the Royal Society. Medical practice rarely allowed Erasmus time to attend meetings of the Society in London, but he exploited fully the personal connections that a Fellowship opened up for him. This he did through letter-writing and by encouraging Fellows to visit him and the other 'Lunaticks' at meetings in the vicinity of Lichfield. There was in his part of the midlands no shortage of inventive, practical men whose interests were as broad ranging as his own. One of the first, whose friendship he made in the early 1760s, was the Birmingham manufacturer and entrepreneur, Matthew Boulton. (It was Boulton and James Watt who later combined their respective practical and inventive geniuses to produce steam engines.)[29] By the mid-1760s, Darwin and Boulton had drawn into their group men such as the potter Josiah Wedgewood I, whose consuming interest was the chemistry of pigments and who, as seen already, was later to be linked to the Darwin family by marriage. Also in the group was Richard Edgeworth, a somewhat eccentric Irish inventor, and William Small, academic, physician, tutor of Thomas Jefferson and friend of Benjamin Franklin. Among the group's more frequent visitors was Joseph Banks, botanist and explorer of the Australasian flora, who in 1778 was President of the Royal Society.[30] Included among those who joined the group later were William Withering, discoverer of the medicinal properties of foxglove, and, most significantly, Joseph Priestley. The last two had very different interests in plants but, as will be seen in Chapter 4, each strongly influenced Erasmus's botanical writings. This celebrated group has become known as the Lunar Society of Birmingham, and its members 'Lunatiks', because its meetings were held on days close to the time of a full moon, when there would be most light to help the participants to travel home in relative safety. Birmingham, then little more than collection of loosely interconnected villages, was where the centre of their activities gravitated in later years; sadly, it proved too far from Derby for the ageing Erasmus to attend their meetings.

Benjamin Franklin, who was keenly interested in the properties of electricity, was only occasionally present at the group's meetings but was a frequent correspondent, while another correspondent was Jean-Jacques Rousseau. Again, it was Erasmus whose friendship established the link with Rousseau, at the time a political exile from France and staying at Wooton Hall in Staffordshire. The story goes that one day in 1766, while Erasmus was walking the Staffordshire moors to Dovedale, he deliberately paused near a cave on the terrace of the Hall, ostensibly to admire a flower. The great philosopher was drawn out of the cave, where it was his habit to sit in melancholic contemplation, to share his interest in plants with Erasmus and, no doubt, to have his spirits lifted by Erasmus's infectious enthusiasm.[31]

Just as Erasmus had loved nothing better than exchanging information and ideas with other philosophers, so too his grandson Charles built up a network through which information was exchanged, with himself at the centre. An obvious difference between the two was that whereas for Erasmus letters were a poor substitute for face-to-face meetings, Charles's preference was to stay at home content to learn through the letters. His letters were always respectful, indeed they were typically kind, as was recalled by Francis his father's secretary through his last years:

> When he had many or long letters to write he would dictate them from a rough copy ... written on the backs of manuscript or proof pages. He was considerate to his correspondents, e.g. when dictating a letter to a foreigner, he hardly ever failed to say to me, 'you'd better try and write well, as it's to a foreigner'. His letters were generally written in the assumption that they would be carefully read; thus, when he was dictating, he was careful to tell me to make an important clause begin with an obvious paragraph, 'to catch the eye'.[32]

Neither Erasmus nor Charles was shy of using their reputations to direct and dominate others, though Charles did this in a much more subtle way than Erasmus.

Charles asked questions of correspondents around the world, always in the most politely phrased way and, usually, he received equally polite answers. After *On the Origin of Species* was published, his fame enabled him to demand even more from his circle of correspondents, which by now included most of the leading Continental biologists and geologists of the age. Sometimes, if they were in Britain, they were invited to Down House, in spite of the fact that Charles found even these meetings, on home territory, were often extremely taxing. Thus, Ernst Haekel, the developmental zoologist, and Ferdinand Kohn, the pioneer microbiologist, was each received – the young and enthusiastic Haekel amused the Darwin children greatly by shouting to their father in broken English.[33]

Charles's fame not only brought the Darwin children into the company of the most important scientists but it sometimes, too, allowed them to meet the most fashionable men and women from the contemporary world of the arts. The range of these contacts was considerably extended through the social circle of their uncle, Erasmus, known as 'Uncle Ras' by his nephews and nieces. (Unfortunately, since it encourages confusion, the Darwin family recycled a few first names over several generations.)

Ras was Charles's beloved elder brother. 'My brother Erasmus possessed a remarkably clear mind, with extensive and diversified tastes and knowledge in literature, art, and even in science'.[34] In boyhood the brothers were inseparable, sharing a passion for collecting plants, insects and rocks. It was Ras who cultivated his younger brother's interest in chemistry and, together, they spent their pocket money setting up a chemistry laboratory in an outhouse of their home in Shrewsbury. When Charles went to study medicine at Edinburgh University, Ras transferred there from Cambridge to complete his own coursework. The brothers' interdependence was such that in the year the two spent together they made no efforts to make new friends, their closeness being re-enforced by an intense dislike of Edinburgh's intellectual and social life. When Ras moved, under their father's direction, to an anatomical school in London, Charles was left isolated in Edinburgh. As he recorded in his autobiography, he had for some time been convinced 'from various small circumstances that my father would leave me property enough to subsist on with some comfort, though I never imagined that I should be so rich a man as I am; but my belief was sufficient to check any strenuous effort to learn medicine'.[35] His dislike of surgery and blood, coupled to his new loneliness, soon convinced him to abandon medicine in favour of the church – an equally unsuitable career – for which he would study at Cambridge.

Meanwhile, Ras found his milieu in London, and in London he remained. Whether through his own ill health, or as a consequence of the well established family habit of disliking medicine after studying it at close quarters, Ras never practised the subject. Instead, he found himself an agreeable home in Great Marlborough Street where he happily settled into the life of a gentleman. It has been suggested by Janet Browne that, 'Erasmus' sexual identity was mysterious, almost hermaphroditic'.[36] There is plenty of evidence in support of her suggestion. One of his closest friends was Harriet Martineau, a popular writer and would-be political reformer. He once joked that she wanted to marry him, but he avoided commitment to this strong-willed woman, as he did to all others. Ras loved being in the company of women, but only in unthreatening circumstances. Typically, he delighted in escorting other men's wives around London society. Jane Carlyle, wife of the historian Thomas Carlyle, was a favourite, but in his affections no one rivalled Fanny Wedgwood (née Mackintosh), his cousin's wife. Four years older than Ras, she was always addressed by him as 'Missus' or 'Wifey'.

Returning in 1836 from nearly four years away travelling with the *Beagle*, Charles had found accommodation close to Ras in London and was naturally included in Ras's social circle. But, as his life was taken over, first by marriage to Emma, then by the move to Downe and, finally, by frequent bouts of sickness, so Charles withdrew from society. It was left to Emma, often aided and abetted by Ras and Fanny Wedgwood, to organize visits from Downe to London; on these family trips to concerts or the theatre there was often a chance for the children to meet members of Ras and Fanny's social circle.

When the children were still small, it was Ras who organized the visit to the Great Exhibition of 1851. In later years, he sometimes accompanied Charles and his family on holiday, particularly enjoying a holiday in 1868 on the Isle of Wight. There the holidaymakers rented a house in Freshwater next to, and owned by, Julia Margaret Cameron, a pioneer photographer whose fame was rapidly spreading. Cameron, who needed the income from her portraits as well that from her property, insisted on photographing Charles and the children. A neighbour and great friend of the Camerons was the Poet Laureate, Alfred Lord Tennyson, whose poetry Emma greatly admired. Staying with Tennyson were the American poet, Thomas Appleton, and his cousin, Henry Longfellow, Professor of Poetry at Harvard University. Charles and Henry were immediately united in their admiration of Asa Gray, the Harvard Professor of Botany, who was a good friend of Charles.

The family connections were widened still further when, in 1872, the Darwins' newly married daughter, Henrietta, moved to Bryanston Street, close to Uncle Ras. Her husband, Richard Litchfield, had his own well established circle of liberal friends which included Arthur Munby, the painter, and John Ruskin, the artist, art-critic and social reformer. Later, in 1879, during a family holiday at the village of Coniston in the Lake District – a holiday which included Francis and the now two-year-old Bernard – Charles met Ruskin who, in spite of being at the height of his fame, lived in comparative seclusion at Brantwood on the opposite shore of Lake Coniston. The two famous men held each other in great mutual respect, and Charles's call was reciprocated. Although, in each man, their love of nature had fundamentally different origins, and Charles always protested his lack of artistic skills, the two shared a deeply held belief that a subject is only fully understood when it is drawn – a valuable principle not forgotten by Francis, the budding biologist.

*****

Last but far from least among the riches of Emma and Charles's children was emotional security. Their five sons and two daughters who reached adulthood received parental love and support in abundance. Charles and Emma encouraged in their children every talent and celebrated every achievement, whether large or small. Charles imposed upon himself a strict and demanding daily schedule of work and exercise and, at least when all too frequent bouts of ill health did not send him to bed, or away for a 'cure', this allowed him little time for play with his children. But he was remembered by his children as being rarely angry, remarkably tolerant of their pranks and misdemeanours, and far from those stern and frightening Victorian fathers of popular fiction. Emma did her best to compensate for the long hours of study that took Charles away from his children, in particular encouraging their musical talents. And then there were the servants, amongst whom Joseph Parslow had perhaps the greatest influence on the growing boys. Employed as butler and Charles's valet, he became a virtual member of the family. He taught the boys to shoot and played billiards with them; he did his best to discipline them – making sure that they took their dirty boots off when entering the house. Parslow travelled with the family on their holidays, interludes when Charles felt free to spend more time with his children, typically introducing them to the wonders of his world of plants, animals and rocks. Francis recollected that Parslow, 'was a kind friend to us all our lives. I do not remember being checked by him except in being turned out of the dining-room when he wanted to lay the table for luncheon, or being stopped in some game which threatened the polish of the sideboard.'[37]

There is a suspicion that the Darwin children may have been a little unruly. Again, Francis recalled,

> I have a faint recollection of black-coated uncles sitting by the fire and not unnaturally objecting to our making short-cuts across their legs. It was no doubt a pity that we were not reproved for our want of consideration for the elderly, and that, generally speaking, our manners were neglected. One of our grown-up cousins was reported to have called our midday dinner 'a violent luncheon', and I do not doubt she was right.[38]

The tight-knit nature of the family was maintained as the children grew to adulthood and was, in turn, passed on to Charles's grandchildren. Friends and those who married into the family recognized a distinctive Darwin character and remarked, 'they were a tight sect and they used "heartening" words ... such as DOLLOP'.[39] Apart from their own language, they shared a peculiar sense of humour and a belief that persistence, intelligence and common sense would inevitably lead to a complete truth.

As well as their shared approach to life, the five sons of Charles and Emma shared a close physical resemblance. Francis's niece, Gwen, confessed

> when I was quite small, the chief difference between them [her five uncles] ... was that three of them had short beards, and the other two had only rudimentary whiskers ... they all had the same family voice – a warm, flexible, very moving voice; the same beautiful hands, and, of course, the same permanently chilly feet.[40]

William, the first of the five boys who survived infancy, was born in 1839, Horace, the last, was born in 1851. They approached adulthood in the mid-Victorian age of the 1850s and 1860s. It was a rare phase in history, for not only was Britain at peace militarily but it was at peace with itself. Both agriculture and industry flourished; prices, wages, rents and profits all went up together. The poor were still there but, in the main, they were deferential. They did not threaten the existing order. It is possible to conclude that mid-Victorian conduct was strongly influenced by two national ideals, 'the gentleman' and 'self-help'. A gentleman could enjoy a safe and solid England and look out on a wider world that was in large part dominated by Britain and British values. The comfortable atmosphere was reflected in the Great Exhibition, which celebrated both the human and the industrial diversity and enormous wealth of the Empire.[41]

Charles and Emma's boys, like their father before them, enjoyed the inestimable privileges of wealth, an enviable position in society, and a good education. If they had spent their lives in idle self-indulgence it would have been no surprise. But, again like their father, they did not. Choosing a variety of careers, each was successful in his own right. William chose banking, while Leonard became a military engineer. Horace too had engineering skills, but his interests turned to small scale machines; he was a founder of the Cambridge Instrument Company, a firm that made precision instruments for the natural sciences, including biology. George and Francis built independent academic careers, George as a mathematician and astronomer, Francis as a plant physiologist.

The many advantages of birth enjoyed by Charles and Francis opened opportunities unavailable to the masses, smoothing their passage through life, but did not make them great botanists. The following chapters will explore exactly what did, examining what they learned from each other, what they took from the wider botanical community, and what their own originality added. And, finally, they will ask, what did they leave behind? Indeed, is their botanical legacy sufficient to justify the accolade 'great botanists'? The exploration starts with Erasmus.

3 THE MISFORTUNES OF BOTANY

In his *Letters on the Elements of Botany Addressed to a Lady* (1771–3), Jean-Jacques Rousseau observed:

> The principal misfortune of Botany is that from its birth it has been looked upon merely as a part of medicine.[1]

Rousseau's diagnosis was right; botany may have been a loosely defined subject in the late eighteenth century but, whatever it was, it was moribund, enslaved by medicine. New knowledge was rarely sought or found. In universities across Europe, botany was taught only to inform prospective doctors how to recognize *materia medica*, plants such as feverfew, foxglove and poppy, of interest for their medicinal properties. It would be another hundred years before botany emerged as a vital, independent science, free from medicine and founded on measurement and experiment.

Similarly, it would not be until the late nineteenth century – when women began to take their place at the laboratory bench – that botany would escape another misfortune; one placed upon its shoulders by male writers, including Rousseau, who, while undoubtedly popularizing the subject, characterized it as an amusing diversion or hobby for gentlewomen.[2] Whereas the poorest of women had traditionally collected from the fields and hedgerows plants for the herbalists, these writers recommended that collecting and drying, naming and drawing plants, were now proper and praiseworthy pursuits for ladies. The problem was not the *feminiz*ation of botany per se but rather its *amateur*ization, and the sorts of botanical pursuits that were being promoted. What hope was there for the subject when the founder and President of the Linnean Society, James Edward Smith, wrote in the Preface to his *Introduction to Physiological and Systematical Botany* (1807):

> In botany all is elegance and delight. No painful, disgusting, or unhealthy experiments or enquiries are to be made.[3]

Botany's development into an independent, experimental science was slow. As late as 1857, the respected literary review, the *Athenaeum*, was still able to complain:

> Of all the natural sciences Botany is perhaps worse treated in this country than any other [because it is] tacked on as an appendix to a course of medical study, and gets little or no consideration in any other direction.[4]

At about this time, in the mid-nineteenth century, botany was suffering yet another misfortune. It was struggling under the weight of a very practical burden, for the trickle of unknown plants that was each year sent back to Britain from its expanding Empire had become a torrent. The tasks of listing, naming and classifying this welter of plants were helped immeasurably by the writings of Linnaeus, translated into English by Erasmus Darwin as described later in this chapter, but the overall effect on botany was stifling. Plants had proved central to the whole enterprise of Empire-building. In the first phase, involving domination of distant lands, there had been the need for the colonists to secure a plentiful supply of food. In the next phase, the emphasis had turned to the growth of cash crops: either, indigenous crops, grown on a scale vastly more extensive than ever dreamed of by the natives, or alien crops, introduced with scant regard to native sensitivities or, too often, to ecological principles. The second half of the eighteenth century was marked by the establishment of botanic gardens, among the earliest being those in the West Indies (1763), Madras (1778) and Calcutta (1786). Their objectives were mixed. They were located at places where an especially interesting local flora could be collected and studied, but any altruism was outweighed by hard headed realism for there was always an underlying recognition that at least some of the plants in the gardens' collections might be exploited for profit. By the mid-nineteenth century, seed companies and rich owners of the largest gardens in Britain had also recognized the commercial potential of the Empire's strangest, most exotic plants, and they were mounting ever more ambitious plant hunting expeditions to Africa, India and China.[5]

The Empire enriched British botany in a quantitative sense, for it increased the number of known and named plants, but in a qualitative sense it impoverished it, for there was precious little time left over for any detailed study of native, British plants. Although nothing could stem the influx of plants, this particular misfortune of botany was to some extent alleviated by Erasmus's popularization of the Linnean system of nomenclature, and his newly defined botanical terminology, for both helped botanists name and classify the plants more efficiently.

At first sight it seems ironic that Botany's liberation from medicine, and the sterility of nineteenth century listing and naming, owes so much to a family, the Darwins, within which there was a strong tradition of studying medicine, the very subject to which botany was in thrall. Both Erasmus, a contemporary and friend of Rousseau, and Erasmus's son, Robert Waring, were practising doctors. In the next generation, Charles began medical training, although he soon gave it up, while his son Francis completed his training although he never practised. As this history of botany unfolds, it will be seen that many of its key players came similarly from a medical background, which should not be altogether surprising because medicine was the only subject open to able young men with an interest in biology and a desire for professional training.

Erasmus's mid-eighteenth century medical training, with its emphasis on recognizing plants as *materia medica* would have clearly impressed on him that the naming of plants was haphazard. All too often one plant bore several names or, potentially more dangerous for any patient taking a plant extract or infusion, the same common name was sometimes given to two or more plants with entirely different properties. To make matters worse, knowledge was not organized because there was no agreed system of classification or hierarchy of categories, a problem that was exacerbated as the flow into Europe of unknown plants from previously unexplored parts of the world accelerated. Confusion was removed and order was brought to chaos only as the writings of the Swede, Carl Linnaeus, percolated through Europe. Beginning with his book *Systema Naturae* in 1735, he advocated replacing the mixture of vernacular names and lengthy Latin 'name-phrases' that were widely used until then with a simple binomial Latin name (a 'proper' name, with genus first and species second) unique to each species. For example, *Lathyrus distoplatyphyllus hirsutus mollis, magno & peramano odoro*,(sweet pea having broad and widely spaced leaves with soft hairs and flowers with a strong and pleasing scent) became simply, *Lathyrus odoratus*. Amongst the other benefits of binomials, errors in translating common names between languages were now avoidable. Linnaeus also proposed that plants should be organized according to the number of male and female organs in their flowers; what came to be known as his 'sexual system of classification'.

The principles of Linneaus's writings soon began to find their way into many English-language pamphlets and books, but herein arose problems. Firstly, only parts of Linnaeus's works were translated, and then often inaccurately with difficult Latin terms, such as names of structures, being carried over unchanged into the English text. Secondly, in setting out the principles of classification, Linnaeus's language was ripe and overtly sexual, to the extent that even some Professors of Medicine were offended. Authors of popular texts, with gentlewomen as a large part of their target audience, had therefore not only to translate but also to bowdlerize Linnaeus. Among those who attempted such a feat was William Withering, a member of the Lunar Society and, at least initially, a good friend of Erasmus. The latter, frustrated because he believed the potential of botany to contribute to the improvement of gardening and agriculture was being held back by the inaccuracies and obfuscations of Withering and others, concluded that Linnaeus's works must be fully and accurately translated from Latin into English. If no one else would do the job then he must. It was an enormous undertaking because each of Linnaeus's books was encyclopaedic, typically approaching 1,000 pages in length. Erasmus wisely tried therefore to enlist help with his enterprise:

his plan was that starting with two neighbours, Brook Boothby and William Jackson, as members, 'A Botanical Society of Lichfield' would be formed and it would translate and publish Linnaeus's works.[6] The new Society did exactly this but, unfortunately for Erasmus, it never acquired more members, and those two that it did have gave him only a limited amount of help. Published at first in instalments, commencing in 1782, and later in 1783 as a complete volume, the first result of his epic undertaking was *A System of Vegetables* (from Linnaeus's *Systema Vegetabilium*). *The Families of Plants* (translated from the last edition of Linnaeus's *Genera Plantarum*) followed in 1787. To be fair, Boothby and Jackson did give Erasmus some help with proof-reading and corrections, and he did receive considerable help from contemporary botanists, most notably from Linnaeus's son and, also, from Sir Joseph Banks, but credit for the two books was due almost entirely to Erasmus. To his gratification, each was well received.

Erasmus probably drew upon one other source of help and that was his brother, [another] Robert Waring Darwin.[7] His brother was by this stage in their lives the 'Squire of Elston', a man of property with a keen interest in gardening. Robert was, like Erasmus, an educated man aware of, and well able to read, Linnaeus's works in their original Latin. Quite independently he had written but not published a manuscript entitled *Principia Botanica* that was intended to be an English language guide to the sexual botany of Linnaeus. It is thought that Erasmus read and learned from this manuscript; the favour was returned however because it was Erasmus who ensured that his brother's work was finally published in 1787. Thus, the two brothers each published major books in the same year, Robert's book selling even better than Erasmus's. When the third edition of Robert's *Principia* appeared in 1810 it contained a dedication to his nephew, Erasmus's son. It read, to 'Robert Waring Darwin, M.D. & FRS ... one so eminent in his profession, by his truly affectionate Uncle, Robert Waring Darwin'. Erasmus had died eight years earlier and it seems likely Robert senior felt he was now free to lavish praise on Robert junior, whose life had not always met with Erasmus's approval, as noted in Chapter 2.

In translating Linnaeus into English, Erasmus did not shy away from sexual language but he was continually faced with the problem of how often should he anglicize descriptive terms? In the event, he reached a happy compromise, retaining many words, such as 'calyx' for flower-cup, but introducing many more new words into the English language of botany that are still standard and commonplace today, words such as 'bract', 'floret' and 'stemless'.[8] Erasmus's translations were, in their turn, paraphrased and bowdlerized by other authors throughout the early part of the nineteenth century, most of whom, but not all, were seeking to popularize botany among women and children.[9]

Writing to his sisters when a schoolboy, Erasmus had often broken into verse, and throughout his life he decorated his letters to friends with lines of poetry. He did not however pluck up sufficient courage to publish any of his poetry until he was fifty-seven years old, in 1789, when his medical reputation was secure. Indeed, it was so high that King George III tried, unsuccessfully, to induce him to move to London to join his team of personal physicians. Erasmus's problem had been that poetry, like botany, was mainly a female pursuit, so to publish a long, light-hearted poem called *The Loves of the Plants* might be seen as unworthy of a serious, professional man. He need not have worried, for after his verses were finally published his medical reputation remained undiminished, while his poetry achieved such popular acclaim that it was deeply admired by the Romantic poets, Coleridge and Shelley.[10]

The popular success of *The Loves* was attributable to its content, particularly the fact that he wrote about plants as if they were people. Here is a typical extract:

> Fair Chunda smiles amid the burning waste,
> Her brow unturban'd, and her zone unbraced;
> *Ten* brother-youths with light umbrellas shade,
> Or fan with busy hands the panting maid;
> Loose wave her locks, disclosing, as they break,
> The rising bosom and averted cheek;
> Clasp'd round her ivory neck with studs of gold
> Flows her thin vest in many a gauzy fold;
> O'er her light limbs the dim transparence plays,
> And the fair form, it seems to hide, betrays.[11]

Chunda is *Desmodium gyrans* (*Hedysarum gyrans*) (Figure 3.1), while *ten* refers to the number of stamens or male organs.

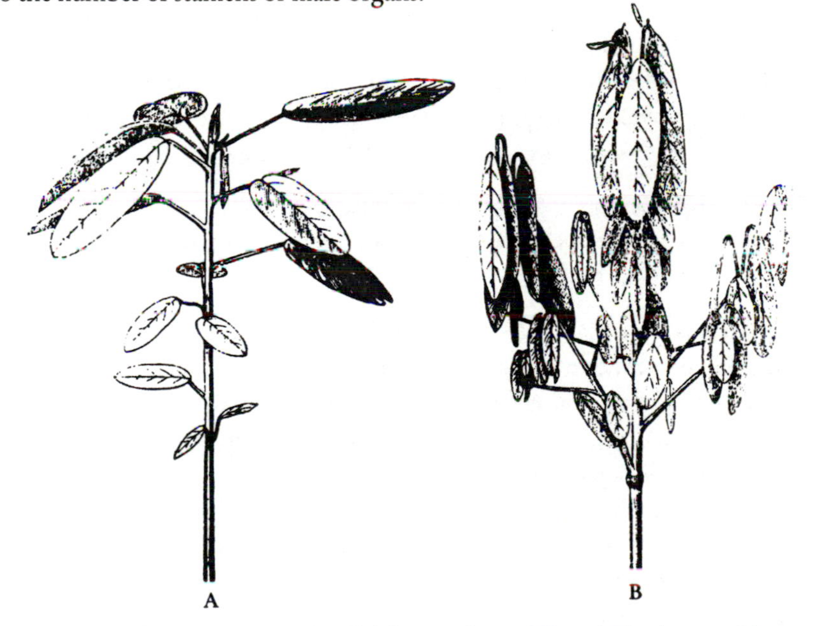

Figure 3.1: *Desmodium gyrans*, commonly called the Semaphore or Telegraph Plant because of the way it raises and lowers its leaves. A, during the day; B, at night with leaves in the 'sleep' position. Reproduced from C. R. Darwin, *The Power of Movements in Plants* (London: John Murray, 1880).

Readers were presented with natural elements in the guise of Gnomes, Sylphs, Nymphs, and Salamanders, and eye-catching chapter headings included, 'Rival Lovers', 'Clandestine Marriage', 'Sympathetic Lovers' and 'Harlots'.

Scholars have debated at length Erasmus's motives in writing *The Loves of the Plants*[12] but there is no doubt that one powerful force motivating him was his very real love of plants, inspired by observations made in his own botanic garden which he had created about a mile west of Lichfield in 1776. It is clear too that he wanted the poem to be read by women as well as men, *The Loves* being designed 'to induce ladies and other unemploy'd scholars to study Botany'. To encourage his female audience, he adopted two subtle stratagems. Firstly, he attracted their attention, but spared their blushes, by using the cloak of botany to mask erotic thoughts and sexual activities and, secondly, he separated as footnotes much of the serious information about plants, which included many of his own observations. His readers could thus enjoy the poem according to their taste, either with or without reading the footnotes.[13]

As he personalized plants, Erasmus had his favourites. Believing that they shared with animals a faculty best described as sensibility – though at a much lower level – he loved insectivorous and mobile plants, which seemed 'almost human'. His fascination with these groups, and his selective focus on a few examples, such as *Drosera* (Sundew) and *Mimosa* (Sensitive Plant), was to prove an inspiration for his grandson, Charles, as will be seen in Chapter 6.[14]

As hinted at in the full title of the poem, *The Loves of the Plants; A Poem With Philosophical Notes*, the footnotes were much longer than the poem. In his notes, Erasmus justified both his selection of plant subjects and their order of presentation in the poem. Not surprisingly, given his commitment to Linnaeus's sexual system of classification, this was the one he chose to follow:

> Linneus [*sic*] has divided the vegetable world into 24 Classes ... distinguished from one another ... by the number, situation, adhesion, or reciprocal proportion [heights] of the males in each flower.[15]

Thus, Class I, One Male, *Monandria*, included plants, such as *Canna*, which possess one stamen, or male organ, in each flower; Class II, *Diandria*, had two stamens, as in *Collinsonia*, and so on. In the case of Chunda, ten was the number of stamens in its flower. Classes were then divided into Orders on the basis of the number of female organs, or styles.

While Erasmus's lexicon and the principles of Linnaeus's system of binomials and hierarchical classification have each withstood the test of time, the principle of Linnaeus's sexual botany, that plants are classified according to their sexual organs alone, has not. As Hunter Dupree, the biographer of Charles Darwin's great friend Asa Gray, politely remarked:

> The Linnean system would classify furniture by placing all four-legged pieces in one group. The natural system would consider the legs as only one of the many characteristics and by weighing several factors place all individuals of a kind, say tables or chairs, into one group.[16]

Rather less politely, the eminent plant physiologist, Julius von Sachs commented:

> It was not Linnaeus's habit to occupy himself with what we should call enquiry: whatever escaped the first critical glance he left alone; it did not occur to him to examine into the causes of the phenomena that interested him; he classified them and had done with it.[17]

A great virtue of the sexual system of Linnaeus, however, was that it was easy to use, particularly by amateurs, which partly explains the willingness with which its principles were adopted throughout Europe, and finally relinquished only reluctantly. Local floras and horticulturalists' catalogues used it, while literate artisans, herbalists and gardeners bought books based on the Linnean system and Latin names. Botany was effectively popularized for ordinary working men, as well as for women, with new botanical societies being formed in places like the fast industrializing county of Lancashire, where their members often met in pubs ('botany at the bottom of a glass') before sallying forth into the field.[18]

For the scientific elite of botany, Linnaeus's sexual system was gradually replaced in the first half of the nineteenth century by more 'natural' systems, although some botanists continued arranging their textbooks and monographs according to the Linnean system well after they had accepted the superiority of more natural classifications. (Thus, William Hooker continued to use the Linnean system in new editions of his *British Flora*[19] until the late 1840s.) The natural systems placed greater emphasis on affinities. Most importantly, they brought attention to small gradations of morphology and, thus, helped prepare the way for evolutionary theory. Among those who led taxonomy out of its cul-de-sac were the de Jussieu's, Bernard and his nephew Antoine Laurent, working in Paris, and, working in French-speaking Geneva, Augustin De Candolle, a man who was to have a marked influence on Charles Darwin's botany, as explained in Chapter 5. They rescued not just taxonomy but botany in general because, under the Linnean system, there was a danger of it remaining entrenched in the pursuit of naming and collecting, the merit of a botanist being judged by the number of species with which he was acquainted. More natural systems opened the way for botanists to think about the interdependence of structure and function, about fitness and adaptations to different environments, and about factors that might limit a species' distribution.

As well as their references to the Linnean system, Erasmus's footnotes to *The Loves* recount folklore associated with plants featuring in the poem and often catalogue their medicinal properties, the latter being familiar to Erasmus by virtue of his medical training and professional experience. Most significantly, Erasmus's notes tell his reader all that was known about what today would be called the physiology and ecology of the plants, referring, for example, to the need of their leaves for oxygen, or the way the timing of bud-burst in tree species reflected spring temperatures in the latitude or altitude at which they grew. In most instances he was recounting what he had been told, or more often had read, but he always gave full credit to the sources of his information, frequently citing the discoveries of friends like Joseph Priestley and Benjamin Franklin. With characteristic honesty, he admitted frankly to what were his own theories and speculations.

Through his remarkable circle of like-minded friends in the Lunar Society, Erasmus was keenly aware that the state of human knowledge was not fixed or finite, that it was continuously growing and edging forwards.[20] It was an age when rapid advances in chemistry, particularly concerning the properties of gases, made it possible for him in *The Loves* often to re-interpret the earlier observations of others on the basic functions of plants. His awareness that new knowledge was accumulating, coupled with his growing feeling that botany should attract serious study, from male scholars, finally drove him to update much of the chemical information he had given in *The Loves of the Plants*, link the notes together in continuous prose format and, shortly before his death in 1800, publish his most important botanical work, *Phytologia or the Philosophy and Agriculture and Gardening*.[21] Where *The Loves of the Plants* can be seen as a product of the Romantic Movement, *Phytologia*, with its emphasis on understanding and improving upon nature, belongs firmly with the Enlightenment, indeed it is a paradigm of that movement. For Erasmus, poetry was meant to amuse while serious matters demanded prose. It is almost as though *The Loves* was testing whether there was an audience for a serious text on botany and, possibly too, testing whether he, Erasmus, was capable of delivering it. If *The Loves* failed then he had no need to go further. But it did not fail and *Phytologia* followed. Erasmus's genius lay in both sensing the changes occurring in the intellectual atmosphere at the end of the eighteenth century and in hastening those changes.

When *The Loves of the Plants* was republished in 1791 as Part II of *The Botanic Garden*, Erasmus offered in his opening lines an 'Apology' to his readers:

> Extravgant theories... in those parts of philosophy, where our knowledge is yet imperfect, are not without their use; as they encourage the execution of laborious experiments, or the investigation of ingenious deductions, to confirm or refute them.[22]

With help from Erasmus, botany was at last starting to escape the shackles of medicine. As will be seen in the next chapter, it was able to move forward in new directions as increasing weight was given to original observation and, especially, to evidence from experiment.

4 ERASMUS DARWIN'S VISION OF THE FUTURE: *PHYTOLOGIA*

With a busy medical practice to service and so many friends and disparate interests, not to mention his own botanic garden to maintain, Erasmus Darwin's busy life left little time for practical investigations on plants. Through his translation into English of the original Latin texts of Linnaeus, and the precision he had given to botanical language, he had managed however to move botany forward in a very practical way. If he had contributed nothing else to botany, he had helped to bring long needed order to existing knowledge. But it was through his book, *Phytologia; or the Philosophy of Agriculture and Gardening*, published just two years before his death, that he made his greatest contributions to the progress of botany.

Its significance is twofold. Firstly, it anticipated the mood of the approaching Victorian age for improvement through greater knowledge. The thrust of much of *Phytologia* is that a better knowledge of plant function can play a key role in improving the growth of crop plants, whether in the garden or the field. There was certainly a desperate need for improvement, for in fourteen of the twenty-three years from 1793 to 1815 there were exceptionally poor harvests in Britain and much of Europe, farmers' problems being exacerbated from 1804 by the Napoleonic wars that engulfed the continent.[1]

Phytologia is significant secondly, because, with its emphasis on measurement and the practical works of the best investigators of the age, it brought together the latest advances in chemistry and botany. It updated what Erasmus had written in *The Loves of Plants*, providing a mature synthesis which – even if it contained as much speculation as it did hard fact – proved seminal, being widely quoted by authors and teachers who followed in the next two or three decades. For the modern reader, *Phytologia* summarizes the state of botany at the start of the nineteenth century. Tellingly, it demonstrates that Erasmus clearly anticipated the emergence from within botany of a separate discipline, the study of function in relation to form. (Ultimately known as plant physiology, the discipline was to dominate all of botany in the later years of the nineteenth and much of the twentieth century.)

The last years of the eighteenth century had seen a significant flurry of activity in the study of plant function (see Table 4.1), with Erasmus knowing either personally, or indirectly via friends, many of the men responsible for the progress that was made. Excepting the works of de Saussure – published immediately after Erasmus's death – and a few others referred to later in this chapter, *Phytologia* drew a picture that was to change little for the next half-century. Remarkably, Erasmus's book was far from outdated when his young grandson, Charles, entered Cambridge University in 1828.

Table 4.1: Early Contributions to an Understanding of the Function of Leaves.

Date	Author	Book	Leaf function(s) proposed
1727	Stephen Hales	*Vegetable Staticks*	Water is lost by 'perspiration' (=transpiration). Air, absorbed through roots, is combined into matter in leaves; he suggested light was necessary for the latter.
1774	Joseph Priestley	*Experiments and Observations on Different Kinds of Air* (first of six volumes)	Leaves restore 'vital air', to air depleted by animal respiration. Requirement for light was not explicit.
1776	Antoine Lavoisier		Brought order to the chemistry of gases, destroying phlogiston theory. Named vital air 'oxygen' and showed that 'fixed air' (CO_2) is a mixture of oxygen and carbon.
1779	Jan Ingen-housz	*Experiments on Vegetables*	Light, not warmth per se, is *necessary* for leaves to make food from the air they capture. In the dark, leaves breathe, like animals, releasing CO_2.
1782	Jean Sénebier	*Mémoires physiochimiques sur l'influence de la lumiere* (3 vols) was the first of four publications, the last being *Physiologie Végétale* in 1800 (5 vols)	Evolution of oxygen is dependent upon a supply of carbon dioxide. Plants fix carbon dioxide in solution as carbonic acid (wrongly, he suggested CO_2 was absorbed via roots)
1800	Erasmus Darwin	*Phytologia*: it was revised and extended from *The Loves of the Plants* (1789).	As water is 'perspired' it is split by light, hydrogen is re-absorbed while oxygen is released. CO_2 is absorbed via the roots. Sugar, which may be converted to starch, is the first product of 'digestion'.
1804	Nicholas T. de Saussure	*Recherches Chimiques sur la Végétation*	In photosynthesis, CO_2 is captured from the atmosphere; water as well as CO_2 is needed.

A period of advance in one science often follows advances in a cognate science, and so it was with botany in the late eighteenth century. Advances in chemistry and physics, most importantly in the understanding of the nature of 'airs' and 'gases', gave Erasmus a new perspective on the plant in relation to its above-ground environment. Practical considerations were, however, largely to dictate that progress remained painfully slow through the first half of the nineteenth century. It is an axiom of modern biology that structure and function are related; without an understanding of structure there can be no proper understanding of function. The problem for Erasmus and his contemporaries was that available microscopes were inadequate for the needs of researchers; the quality of their lenses was not good enough to resolve the details of structures of interest. Anatomists were limited to (imperfect) studies of tissues and organs rather than of individual cells and their interconnections. There were conceptual problems too, for another half-century would pass before – with the help of improved microscopes – Matthias Schleiden[2] and Theodor Schwann,[3] respectively botanist and zoologist, first formally articulated 'The Cell Theory', announcing that the cell is the basic and universal building block of *all* living organisms. (The Germans acknowledged their debt to Robert Brown, discoverer of the cell nucleus, whose friendship with Charles Darwin is described in the next chapter.) When Schleiden's microscopy soon proved that sexual reproduction occurs in non-flowering plants, such as mosses and ferns, as well as in flowering plants, he ushered in a period when large strides forward were taken in the understanding of sexual reproduction in plants – it was the dawning of what might be called sexual physiology, as opposed to vegetative, or growth, physiology to which the Darwins were to bring light.[4]

What had puzzled men for centuries was the ability of plants to gain weight, and volume, i.e. to grow, without feeding on or ingesting solid materials as animals did. The soil seemed the obvious source of nutriment, but the means by which nutrients were acquired or interconverted was a complete mystery. Rejecting the prevailing, centuries old, Aristotelian doctrine of the four elements and plants obtaining 'predigested' nutriment from the soil, the Belgian physician and chemist, Van Helmont, had in 1648 proposed that the whole substance of plants, the mineral parts (the ash) as well as the combustible organic parts, were formed from water. To prove this, in what was probably the first scientific experiment ever carried out on a plant, he had planted a willow tree in a weighed amount of earth. After watering it faithfully for five years, he had found that while the tree had gained 164 pounds in fresh weight, the soil had lost only two ounces in weight (1 pound = 16 ounces). This proved to him that the tree represented water transformed into another substance. In the alchemist's terms, water, which contained the principles of mercury and sulphur, was transmuted into earth. Van Helmont's experiment was later repeated, somewhat more elegantly, by Robert Boyle[5] who similarly concluded that water had been transmuted into the various substances of the plant body. Both men were confused victims of the alchemical dogma of their age and neither had distinguished fresh weight from dry weight, so missing the critical fact that water is responsible for over 90 per cent of plant fresh weight (of the remaining 10 per cent, usually less than 10 per cent of that is made up of mineral elements; the bulk is made up of organic compounds derived from photosynthesis). However, it was here, with water, that experimental plant physiology and the quest to explain growth began.

From the outset it was seen that water is fundamental to plant life. The next step was to connect transpiration, the loss of water from aerial surfaces, with the process with photosynthesis, the capture of carbon dioxide by leaves. The two processes are intimately related because, as Francis Darwin was to discover, stomatal pores in leaf surfaces and green stems exhibit daily cycles of opening and closure through which they simultaneously regulate the efflux of water and the influx of CO_2. Knowledge of transpiration facilitated an understanding of photosynthesis, and in time the complement was returned.

$$CO_2 + H_2O + \text{light} = O_2 + \text{organic matter (photosynthesis)}$$

Plants, like animals, lose dry weight as they release carbon dioxide back into the atmosphere through either their respiration (below), which continues in both light and dark, or their combustion, either before or after they are fossilized:

$$O_2 + \text{organic matter} + \text{light or dark} = CO_2 + H_2O \text{ (respiration)}$$

Apart from its role as a solvent for chemical reactions, and as an occasional reactant itself, water is necessary also for the expansion of cells. It is the turgidity of tissues that enables leaves to be erect and small herbaceous plants to hold themselves upright on land. Loss of water is a price terrestrial plants have to pay for the acquisition of carbon. How to minimize that water loss while maintaining photosynthesis has been a constant evolutionary challenge for plants ever since they migrated out of the primeval oceans, into swamps and finally onto the land. (As will be seen in the final chapter, a challenge that confronts modern plant physiologists is how to *help* plants conserve whatever water is available in the soils of fields or forests.) So it is with the questions surrounding how plants gain and lose water that studies of plant function and growth began.

*****

Erasmus Darwin's medical education began in earnest in 1754 when, after three sterile and uninspiring years in Cambridge and then London, he moved to Edinburgh to continue his studies.[6] In that cold, grey Scottish city he was warmed by professors who were the most enlightened and advanced in Britain. The physics of Isaac Newton, albeit adapted for the human body by Herman Boerhaave in Leyden, may still have predominated, with their emphasis on pipes, pressures and plumbing, but newer ideas from the continent were beginning to infiltrate the dogma. In Erasmus's education in Edinburgh there was emphasis on the nervous system, on the integration of nerve and muscle, and a new appreciation that ill health could arise from external factors, not just fractures or blockages of the pipework. It is not surprising therefore that when Erasmus eventually turned his attention to plants, he was inspired by the Reverend Hales, a man whose approach to plants was entirely Newtonian.

Phytologia is long and at times rambling. Both its full title and its dedication, 'to Sir John Sinclair, Baronet, to whose unremitting exertions, when President of the Board of Agriculture, many important improvements in the cultivation of the earth were accomplished and recorded ...', reveal Erasmus's intention to provide a scientific basis for the improvement of plants and their cultivation. Implicit in his thinking was his belief that species had improved in the course of natural evolution and, therefore, might be improved further by man, but *Phytologia* is not a manual of plant breeding. Rather, after a first part, 'The Physiology of Vegetation', devoted to plant anatomy and function, the second part is given over to 'The Economy of Vegetation' and contains chapters on such practical matters as 'Manures or the food of plants' and 'Aeration and pulverisation of the soil'. Part three, 'Agriculture and Horticulture', describes methods for the production of fruits, seeds, and bulbs. It ends, rather disconnectedly, with a distantly related subject, a 'Plan for disposing of a part of the system of Linnaeus into more natural classes and orders', clearly showing Erasmus was well aware of the limitations of the Linnaean system. A spirit of improvement permeates the whole book. Its Introduction begins:

> Our imperfect acquaintance with the physiology and economy of vegetation is the principal cause of the great immaturity of our knowledge of agriculture and gardening. I shall therefore first attempt a theory of vegetation, deduced principally from the experiments of Hales, Grew, Malpighi, Bonnet, Du Hamel, Buffon, Spallanzani, Priestley and the philosophers of the Linnean School, with a few observations and opinions of my own ... Some notes from *Zoonomia* or *Botanic Garden* are enlarged upon here.[7]

His core belief was that vegetables [plants] are an inferior order of animals. This belief appears to stem from the observation, mentioned several times, that:

> If a bud be torn from the branch of a tree or cut out and planted in the earth with a glass cup inverted over it, to prevent exhalation being greater than its powers of absorption, or if it be inserted into the bark of another tree, it will grow and become a plant in every respect like its parent.[8]

Such an observation proved to him that every bud was an individual vegetable being. A tree was a family or swarm of individual plants – rather like a coral polyp.

If plants were merely inferior forms of animals, then he could explain their functions in familiar terms. More was known about animals than about plants so this approach was usually helpful, but in some areas Erasmus pushed the analogy between animals and plants far beyond the point at which it stopped being helpful. Thus, while accepting that vegetables differ from animals in that they do not have muscles of locomotion or digestion, he believed vegetables contain longitudinal muscles to turn their leaves to the light and to expand or close their petals or sepals. They also possess, he thought, vascular muscles to perform the absorption and circulation of their fluids, with the attendant nerves, and a brain, or a common sensorium, belonging to each seed or bud. He proposed that plants have three types of absorbent vessel: the first imbibes the nutritious moisture of the earth, just as the lacteals absorb the chyle from the stomach and intestines of animals; the second imbibe the water of the atmosphere, opening their mouths on the outside of leaves and branches, like the cutaneous lymphatic system of an animal; and the third imbibe fluids secreted from the internal cavities of the vegetable system, like the cellular lymphatics of animals.

Such understandable minuses are far outweighed by the positives in Erasmus's writing about plant function. His starting point was Stephen Hales (1677–1761), who can justifiably claim to be the founder of plant physiology.[9] Oddly, Hales would have been confused by that claim for, in his day, when the word 'physiology' was used in a botanical context it meant all aspects of the study of plants – except systematics. It was not for another hundred years, roughly halfway through the nineteenth century, that the term 'experimental physiology' began to be used to distinguish a unique discipline within botany, and still later in the century that 'physiology' assumed its modern meaning, which is the study of the functioning of organs, or the functional inter-relations between organs, in relation to their form.

Hales published just one book on plants, *Vegetable Staticks* (1727). The key to understanding why it was so important lies in the word 'Staticks'. Inspired by Isaac Newton and Robert Boyle, he did something quite revolutionary. He not only experimented with plants but, by weighing and measuring them, he quantified his results. From Hales onward, precision mattered. At the core of his 'statical way of enquiry' there was an underlying assumption that there was a functional equilibrium within the organism, requiring measurements of both inputs and outputs so that any inequalities might be accounted for.[10] Hales is as important for *how* he found out as for *what* he found out.

In conformity with his strong religious faith, he believed that an all-wise Creator had observed the most exact proportions, of number, weight and measure.

To reveal that pattern was, therefore, to reveal the wonder of God's works. In the Introduction to *Vegetable Staticks* Hales wrote:

> in relation to those Planets which revolve about our Sun, the great Philosopher of our age [Newton] has, by numbering and measuring, discovered the exact proportions that are observed in their periodical revolutions and distances from their common centres of motion and gravity: And that God has not only comprehended the dust of the earth in a measure, and weighed the mountains in scales, and the hills in a balance, Isai. Xl 12. but that he also holds the vast revolving Globes, of this our solar System, most exactly poised on the common centre of gravity.[11]

Erasmus and Hales may have differed fundamentally in their belief in God – Erasmus had none – but for both men an interest in the physiology of plants sprang from an earlier interest in the physiology of humans and animals. Hales was educated at Corpus Christi College (then called St Benedict's), Cambridge, and was always destined for the priesthood. However, while still a Fellow of the College, an interest in physiology was kindled by his friendship with a medical student, William Stukely, and by James Keill's lectures in [human] physiology and anatomy. With the discoveries in mind of another Cambridge man, William Harvey, who in 1628 had demonstrated the pressurized circulation of blood in the human body, Hales's first investigations, during which he perfected his statical approach, were therefore conducted on the circulation and pressure of blood in animals. In what was among the curate of Teddington's more (in)famous experiments, he inserted brass pipes or glass tubes into the arteries of horses tied either on their back or side-on to a field gate. His objective was to measure their blood pressure. These experiments on domestic animals such as horses and dogs are brutal to modern eyes – they even shocked some of his contemporaries, including his friend and neighbour Alexander Pope – but in spite of the horses' struggles he was able to make measurements that compare well with modern ones.

> Green Teddington's serene retreat,
> Where he good Pastor Stephen Hales
> Weighed moisture in a pair of scales
> To lingering death put Mares and Dogs,
> And stripped the Skins from living Frogs.
> Nature he loved, her Works intent
> To search or sometimes to torment.[12]

His account of this work, in *Haemastaticks*, was not published until 1733, some six years after *Vegetable Staticks*. What induced him to turn his attention to plants is not clear – possibly he was affected by criticisms of his experiments – nor does he reveal how long it took him to adapt his techniques to plants. The switch from animals would not have been straightforward but it was effected successfully and he was confidently into his statical stride in the very first chapter of *Vegetable Staticks*, 'Experiments Showing Quantities Imbibed and Perspired by Plants and Trees', which exemplifies his radically new approach. He used sunflowers, cabbages, spearmint and a number of woody plants ranging from vines to apples and laurel. In the first experiments, Hales made repeated measurements on a 3 ½-ft-high sunflower planted in an unglazed pot, with the soil covered with a thin lead sheet to prevent evaporation. Hales estimated that in four cubic feet of watered soil there was sufficient water for twenty-one days' growth of his sunflower. Over fifteen days there was only a small change in plant fresh weight. At the end of the period, he cut off the plant and measured weight lost through the sides of the pot. Then, knowing the amount of water that he had added each night and morning, and subtracting an amount of water lost through the sides of the pot, he calculated the difference, which was the amount of water 'perspired', or transpired, by the plant. He found perspiration was greatest in the middle of warm, dry days, and was least in a cool, moist night (see Table 4.1).[13]

After calculating the surface area of leaves, stem and roots, he calculated the 'perspiration rate', that is the volume transported across each unit of area (inward fluxes for roots, longitudinal or axial fluxes in the stem, and outward fluxes for aerial parts of the plant). Rates were highest in the stem, which for Hales was analogous to the arteries and veins of animals.

Another aspect of the genius of the man, and one that further endeared Hales to Erasmus and his generation, was that he always attempted to relate his measurements to the known behaviour of plants in the field and to agricultural practice. Thus, he compared measured fluxes for the same species growing under different climatic conditions, and for different species in the same environment. He noted, for example, leaves of evergreen species lost water more slowly than those of deciduous species and, so, he was able to offer an explanation of the former's survival through winter.

A desire to compare plants with animals was probably one of the factors that drove him to turn his attention from animals to plants. When he compared the velocity of sap movement in stems with that of blood through veins and arteries, using values for perspiration given by James Keill in *Medicina Statica Britannica*, he concluded that the volume of fluids passing through plants is considerably more than that passed through man.[14]

But what forces, he pondered, were responsible for the flow of sap? Was capillarity responsible? In a series of ever more complex experiments, Hales confirmed that capillarity did indeed cause sap to rise through tubes of extremely narrow diameter, but the final height attained came nowhere near matching the height above ground of leaves on tall plants which obviously had no difficulties in obtaining sufficient water.

Hales found it was the leaves themselves that determined the volume of sap that flowed in a given time. As proof of this, in his simplest experiment, he cut pairs of branches, each of similar size and shape, from several varieties of fruit tree. From one of each pair, cut from the same tree, he stripped off all the leaves, and then set each branch in a separate vessel containing a measured quantity of water. The branches with leaves imbibed 15 to 30 oz of water during the night: those without leaves took up only approximately 1 oz. In a subsequent experiment, he cut a branch from a tree and attached a 7 ft long glass tube to it; hanging the branch and tube upside down and filling the tube with water, i.e. with the tube uppermost, he found that the water remained at the same height in the tube if the branch was immersed in water, but was drawn down the tube if the branch was exposed to air (into which it perspired).

Some botanists held that, like blood in the human body, sap circulated around the body of a plant, in trees travelling up in the older wood and down in the younger wood. Hales showed clearly, and for the first time, water will travel either way through excised sections of woody stem, the direction being dictated by the driving force, i.e. it is fully reversible. When he cut deep holes in stems, arranged alternately to interrupt the ascending sap vessels, the mercury manometer showed water was still imbibed. This demonstration of lateral transfer, together with the previous demonstration of reversed flow, argued very strongly against there being any circulation of the sap *within* the wood of trees (a little water does move downwards in tissues *outside* the wood, carrying the dissolved products of photosynthesis, and that water is later recycled upwards). Although Hales's starting point was that animal and plant systems were analogous, his observations from experiments gave him the courage to proclaim differences where he concluded they occurred.

To round-off his brilliant analysis of the causes of sap flow, Hales considered the possibility that a previously unrecognized factor, root pressure, might be involved. What he noted was that vine stumps and some trees bled when cut at certain times of the year, e.g. in Spring *before* the leaves opened. He attached tubes to the stumps and measured the volume of exudate, which was particularly copious during warm daylight hours. When he attached a mercury manometer he saw that a pressure equivalent to 43 ft of water occurred (seven times that in the crural artery of a dog). However, if he attached manometers to different branches of the same tree simultaneously, he found different rates of exudation – and occasionally even absorption – confirming his belief that a pull-from-above, from transpiring branches, is the dominant force driving the ascent of sap in intact plants. This proposition – arguably his most important – has often been challenged but never refuted in nearly three centuries that have followed.

It is to underestimate Hales to suggest that he contributed to, or advanced our knowledge of, plant water relations because the truth is that before him there was nothing. There were no great men upon whose shoulders he could stand. He not only laid the foundations but he built upon them a considerable edifice. His achievements were enormous, although of course he was occasionally wrong and misled his successors. For example, in his first experiment on the bleeding vine Hales had observed 'a continuous series of bubbles constantly ascending from the stem through the sap in the tube'. He thought this was a sign that air entered through the roots, a misconception only strengthened when, cementing the stems of cut branches into closed vessels of air, he found that air was drawn up as water was 'perspired off'. He did not realize that opening the conducting vessels to the air is unnatural – each vessel is normally sealed to exclude air because bubbles (embolisms) block water movement – and led, therefore, to misleading results. Similarly, he did not realize that in intact roots the endodermal cylinder stops air crossing the root radially, so protecting the vascular system which it ensheaths from the ingress of air. He did not know something which Erasmus was to discover, which is that only part of the cross section of the stem (the xylem tissue, which includes numerous vessels) transports water. Nor did he know that most water transpired from leaves and green tissues of the shoot is lost through specialized epidermal pores, or stomata, the same pores through which oxygen enters and carbon dioxide both enters and exits the leaf. In all this it should be remembered that Hales's concept of 'air' was very different from our own.

Quite naturally, Hales was a victim of his times, confused about the nature of 'air'. He had to struggle under the early eighteenth-century misconception that air was a fine elastic fluid, with particles of very different natures floating in it. The lengthy and convoluted title of the longest chapter in *Vegetable Staticks* indicates the confusion reigning in Hales's mind; the chapter is entitled, 'A Specimen of an Attempt to Analyse the Air by Chymio-statical Experiments, Which Shew, in How Great Proportion Air is Wrought into the Composition of Animal, Vegetable and Mineral Substances: and withal, How Readily it Resumes its Elastick State, When Dissolution of those Substances it is Disengaged from Them'. Changes in the volume of air were explained as being due to changes in elasticity. In combustion, or respiration, the elasticity of the air was destroyed by the addition of unelastic vapours, until the air was so saturated with absorbing particles that its activity had to stop. In spite of such confusion, and the analogy he drew between leaves and lungs, Hales left for Erasmus and those who came after him one indelible truth; plants use 'air' for growth (see Table 4.1); they are not made from water as Van Helmont had proposed.

Hales isolated various 'airs' or gases but he failed to recognize their distinctive chemical properties. Nevertheless, even here his work was not without value. His concept of 'fixed air' was widely influential. It would be proved that this air, now known as carbon dioxide, could be incorporated into the substance of plants, animals and minerals, and could also be released. Erasmus never met Hales, who died in 1761, so there was no direct connection between the two men but he did know very well the Reverend Joseph Priestley who, although late to join, became a leading member of the Lunar Society.[15] Erasmus frequently quoted the chemical and botanical discoveries of his friend and, occasionally pointed Priestley's researches in the right direction.

Priestley was, like Erasmus, fond of quoting Hales but more importantly he sometimes used Hales's reports as a starting point for his own investigations, as for example in his isolation of nitrous oxide (laughing gas, N_2O). Although both men were Christian ministers, with deep concerns for the poor and strong views against the consumption of alcohol, the two could hardly have differed more in background or personality. Whereas Hales came from wealthy land-owning stock, was educated at Cambridge and, at least in his social life, was typically a conformist patrician, Priestley came from poorer stock. He was a restless plebian whose academic training, at the Daventry Academy, was aimed at fitting him for life as a dissenting minister.[16] The son of a Yorkshire cloth-dresser, he had been brought up in straightened circumstances by a strongly principled Presbyterian aunt after his mother had died when he was only six years old. Recognizing the boy's latent abilities, a local minister had taught him classics and he learned several European languages. Although he was a fervent and dedicated non-Conformist minister, experimental chemistry was his hobby, a hobby he was increasingly able to pursue as his fortune and reputation increased. Science was for him a vehicle by which man might reach the ultimate happiness intended by Providence. His religious principles were mixed with socialist principles; indeed, so strident were his views that in 1791 he was driven to flee Britain for Pennsylvania after a mob destroyed his home, laboratory and Meeting House in Birmingham angered by their mistaken belief that he was trying to import the French Revolution.

Long before his hurried departure, Priestley had written in 1770 that he was about to take up 'Dr Hales's inquiries concerning air'. Priestley's use of such expressions as 'releasing' air, 'diminishing' it, or 'changing its elasticity', and his persistent dependence on volumetric parameters, all reveal the Halesian origin of, and influence on, his basic pneumatic concepts.[17] Hence it was consistent with Hales's philosophy when Priestley found that a candle burning in air in a closed system was extinguished when the volume decreased by twenty percent – roughly the fraction of oxygen in air. (Priestley's apparatus amounted to a simple eudiometer – a device for collecting gases and measuring their volume.) More than pneumatic changes had happened, however, for the air was in some way 'injured' so that a mouse placed in it would soon swoon and die.

Priestley's investigations did not always proceed in the orderly fashion that one expects of a famous scientist. Thus, his next step, his great discovery in 1771 that plants had the power to restore injured or putrid air (see Table 4.1), was made in what another Lunar Man, James Watt, described as 'his usual way of Groping about'.[18] It seems Priestley put a sprig of mint into a glass jar containing air in which two mice had died and forgot about it. The mint was standing in water and, to Priestley's surprise, not only continued to grow but made the air fit for a new mouse to live in.[19] He may have stumbled towards the discovery but with the same genius that was possessed by Charles Darwin for separating the single significant event from a multitude of trivial ones, Priestley suspected something of the greatest importance had happened. His suspicion was confirmed when he found a candle would now burn in the jar and a mouse could breathe freely in it. The plant had restored the air, it had produced 'dephlogisticated air' (oxygen). He immediately tried the same with other plants, finding that spinach had the most restorative effects. Though in the next few years he struggled to repeat his experimental findings with any consistency, the positive news that Priestley had discovered the properties of what would prove to be one of the basic components of photosynthesis was brought to the Royal Society in 1772 by Benjamin Franklin.

The term 'dephlogisticated' was applied to the new air because it was believed that all substances contained phlogiston, which escaped when they burned. The more phlogiston they had, the more fiercely they burned. True, air was needed for burning, but this was because it provided a medium into which phlogiston could escape. No one could see or hold phlogiston, which was sometimes regarded as a force rather than a substance. To explain the *gain* in weight that occurred when a metal was burned – when an oxide formed – it was even suggested that phlogiston had a negative weight.

It was not until August 1774, when he heated red mercury by means of a burning glass, that Priestley produced the same dephlogisticated air by *chemical* means. Realizing the importance of his discovery, he promptly announced in volume one of his *Experiments and Observations on Different Kinds of Air* (1774) that he had isolated 'vital air' (= dephlogisticated air = oxygen). A Swedish apothecary's assistant, Karl Scheele, had isolated the same gas, or 'fire air' as he called it, three years before Priestley but, for various reasons including a delay in publication, his discovery was not made public until after Priestley's announcement. Priestley has been enshrined as the 'discoverer of oxygen' not just because of his precedence in publication but because he was better connected with the other leading scientists of his day, with the result that they fed on his discovery, not on poor Scheele's. Thus, by October 1774 he was in Paris telling Antoine Lavoisier, perhaps the most famous chemist of the age, about his discovery.

It was Lavoisier who gave the name 'oxygen' to Priestley's air and who, in 1776, recognized that other essential component of photosynthesis, carbon dioxide or 'fixed air' (see Table 4.1). Our modern system of chemical nomenclature is attributable to this wealthy Frenchman, whose wife translated for him the works of Hales as well as those of Priestley and whose partnership in the private agency that collected taxes for the government was in 1794 to cause him to lose his head in the French Revolution. Like Linnaeus, Lavoisier had a genius for organization. He brought clarity where there was confusion. For example, a single name, oxygen, replaced three old ones, 'fire air', 'vital air' and 'dephlogisticated air'. Lavoisier's new terminology clearly indicated that respiration in animals, like the combustion of organic materials, was an oxidative process, i.e. one in which atmospheric oxygen was consumed as it combined with other elements. For Lavoisier it was simple, both respiration and combustion comprised the absorption of oxygen and release of carbon dioxide and heat energy.

Most of the chemists in England, including Priestley, continued to believe in phlogiston throughout the 1780s but, remarkably, Erasmus Darwin was the first Lunar man to be converted to Lavoisier's views and terminology. It was Erasmus too who, via James Watt, convinced Priestley to prove by experiment that water was not itself an element but was composed of hydrogen and oxygen – thus confirming, Erasmus thought, his view of leaf function. Gradually, discoveries in chemistry opened the way for a proper understanding of the processes of respiration and photosynthesis.

One of the very few men of the age whose talents were as great and awesomely broad as Erasmus Darwin's, arguably more so since as diplomat, envoy to Europe and politician he became a father of the American Revolution, was Benjamin Franklin. The two probably first met in 1758,[20] although they were already aware of each other's interests thanks to their connections with Royal Society. Indeed, one of the many talents they shared was for making connections – between people as well as between ideas. It is in character, therefore, that it was Franklin who, in his tour of northern England in 1771, introduced Darwin[21] and, also, Joseph Priestley, to his travelling companion and good friend, Jan Ingen-Housz.[22] The Dutch physician was about to become the next key figure in uncovering the basic properties of photosynthesis, in particular its requirement for light.[23]

The difficulty Priestley had experienced in repeating his demonstration that plants restored (or re-oxygenated) putrid air is explained by his failure to recognize the importance of light to the photosynthetic process. Many of his attempted repeats seem to have been conducted indoors in December when, even at midday outdoors, there is insufficient light for measurable photosynthesis in most plants. Such difficulties make it all the more creditable that Hales should have written so perceptively,

> And may not light also, by freely entering the expanded surfaces of leaves and flowers, contribute much to the ennobling the principles of vegetables; for Sir Isaac Newton puts it as a very probably query, 'Are not gross bodies and light convertible into one another? And may not bodies receive much of their activity from the particles of light, and of light into bodies, is very conformable to the course of nature, which seems delighted with transmutations'.[24]

In an age of unorthodox life histories, Ingen-Housz's manages to stand out. His introduction to Franklin, and in turn to Erasmus, was made possible by virtue of his earlier friendship with Sir John Pringle, President of the Royal Society. They had met when Pringle was an army physician stationed in the south Netherlands during the war of the Austrian Succession. Pringle became friendly first with Jan's father, who was a prominent burger in Breda, and then with his eldest son, Jan. The young man became a physician too but his hobbies were electricity and chemistry and he corresponded regularly with Pringle, who had similar interests. In 1765, Pringle enticed Jan to work in England, where he introduced him to Franklin, who had similarly just arrived, and to Priestley. Despite the differences in character between the intense, passionate Priestley, the exceptionally relaxed and good natured Franklin, and the young unassuming Dutchman, all shared an intense interest in electricity, the list of whose properties was lengthening by the day. Lasting friendships were forged and mutual respect quickly developed. Even several years later when the interests of Priestley and Ingen-Housz had moved on to gas exchange in plants, and they were in dispute over the precedence of their discoveries, their relationship remained respectful

Ingen-Housz won fame in the winter of 1767–8 not for his work on plants but as a physician. With others, particularly the English physician Thomas Dimsdale, he improved a centuries-old technique (variolation) for inoculating against smallpox, something which may have specially endeared him to Erasmus whose face was badly scarred by a smallpox infection suffered in childhood. The method was very risky since it used a live vaccine, and relied heavily on the skills of the physician, but it was a lot better than nothing and, until Edward Jenner's discovery in 1796 that cowpox vaccine had the same protective effects but carried little risk, it was widely adopted during the next thirty years. Ingen-Housz's skills were well known to Pringle, who by this time was a personal physician to King George III, so when Empress Maria Theresa of Austria was desperately searching for an escape from the smallpox which had disfigured her and killed several of her children, the King commended Ingen-Housz to her. He took up the imperial appointment in 1768, successfully vaccinating three of the royal children, and carried his inoculation technique to the corners of the Austrian Empire. The relevance of Ingen-Housz's court appointment is that it provided him with both money to travel and a certain cachet that made introductions easier. Thus, through the 1770s Ingen-Housz made journeys to France and to England (where in 1771 he was made a Fellow of the Royal Society) and in this way he was able to keep abreast of the works of Lavoisier, of his new friend Priestley, and of other leading researchers, such as the physicist, Felice Fontana, who, fortuitously, described to Ingen-Housz his new improved eudiometer.[25]

The court in Vienna made heavy demands upon his time so when he was invited to London to deliver the Royal Society's prestigious Bakerian Lecture in June 1779 he took the opportunity to extend his escape from drudgery. He rented a large and comfortable house at Southall Green, a day's carriage ride from central London, and there he embarked on what would amount to over

500 experiments on gas exchange in plants. Employing Fontana's eudiometer, a device for collecting gases, he experimented on detached leaves immersed in water. When illuminated these would liberate oxygen in bubbles, which could be conveniently and accurately measured. His experiments went well and by the autumn of 1779 he was able to draw his conclusions together. In *Experiments upon Vegetables, Discovering their Great Power of Purifying the Common Air in the Sun-shine, and of Injuring it in the Shade and at Night*, Ingen-Housz stated clearly and correctly that plants derive their sap from the bottom [roots], but their food substances mainly from the air; that the capture of these food substances from the air takes place in the green leaves and depends on the amount of sun*light* and not on the warmth of the sun. Ingen-Housz declared that plants as well as animals breathe, in which process carbon dioxide is given off by all parts of the plant. And plants give off from both green and non-green tissues a gas [oxygen] useful for the maintenance of animal life, while animals give off a gas [carbon dioxide] that serves as food for plants (see Table 4.1).[26]

Before returning to Vienna, Ingen-Housz visited Paris where he personally delivered a copy of his book to Franklin (who included Lavoisier in his closest circle of scientific friends and correspondents). Ingen-Housz also, and more significantly, went to Bowood House, Wiltshire, at the invitation of William, first Marquess of Lansdowne. The Marquess – an occasional patient of Dr Robert Darwin[27] – regularly brought together in his large country house the leaders of fashion and philosophy. He was, in particular, a patron the sciences and had since 1773 provided at Bowood a small, well equipped chemical laboratory for Priestley, who was nominally employed as tutor to the Marquess's children. It was at Bowood that Priestley had 'discovered' oxygen in 1774 and it was here in1779 that Ingen-Housz gave to Priestley a copy of his book.[28] (It was at Bowood too that in 1799 Ingen-Housz, by now financially ruined, died secure in the protection of the Marquess; he is buried in nearby Calne).

Ingen-Housz gratefully acknowledged what he had learned from Priestley but, sadly, although Priestley admitted in private that he was 'much pleased' with Ingen-Housz's book, there was little reciprocation in public. For example, in Priestley's 1790 volume of *Experiments and Observations on Different Kinds of Air*, Ingen-Housz's book is mentioned in the text merely as an incomplete footnote and Ingen-Housz's name is omitted from the index. The result was that, since Priestley was by 1790 a very well known and respected figure in the scientific establishment, Ingen-Housz did not in his lifetime receive the credit he deserved. In contrast, Priestley was, and often still is, celebrated as the 'discoverer' of photosynthesis.

To make the situation even more hurtful, Ingen-Housz became involved in a dispute with another researcher, Jean Sénebier, about the precedence of his discoveries. It seems Sénebier was aware of Ingen-Housz's researches while they were still being carried out, and he may even have copied some of his experiments. Sénebier had been trained as a chemist (unlike Ingen-Housz), although he became a clergyman and later librarian in Geneva, where his Protestant family had fled to escape religious intolerance in their native France. When Sénebier published his own findings in 1782[29] it was clear that in emphasizing the absolute requirement for light in photosynthesis he agreed with Ingen-Housz, but in two key aspects they differed. In the first, Sénebier argued, incorrectly, that the carbon dioxide which was to be the food of the plant was absorbed with water from the soil and then transported to the leaves, while Ingen-Housz realized correctly that it came directly from the air. In the second, Sénebier correctly recognized that the evolution of oxygen by green tissues was absolutely dependent on a supply of carbon dioxide, a fact not appreciated by Ingen-Housz (see Table 4.1). Sénebier's book had its faults and may have been in von Sachs's words, 'tediously prolix', full of 'tiresome displays of rhetoric',[30] but its author was a careful experimenter, conscious of the need to employ healthy plant material and to employ standard conditions during experiments. His long-winded writing provided a valuable second witness to the discoveries of Ingen-Housz and extended them in at least one important aspect.

There is a later botanical paper of Ingen-Housz, published in 1796 in an obscure report, which has until recently been overlooked.[31] In it he not only reviewed his 1779 paper, this time updating it with Lavoisier's new terminology, but he clearly recognized for the first time an important phenomenon that would not be fully understood until the birth of microbiology, some hundred years later, namely, soils (in practice, the respiration of soil microbes) generate CO_2 while consuming O_2. Incidentally, in the same paper, he praises the verses of Dr Darwin.

*****

To return to Erasmus Darwin, what was the value of *Phytologia*, a book in which his own original observations occupy only a very small part? The simple answer is that its value was considerable, for it collected and reported at length the experiments and conclusions of those, from Hales to Sénebier, who had in the previous hundred years made practical investigations into the functioning of plants. The fact that Erasmus could add only a few original observations of his own meant he could play the part of the objective observer when drawing together the work of others. Where drawing inferences from their work, and particularly where speculating upon their implications, *Phytologia* is an entertaining cocktail of half-truths, good sense and inspiration.

Thus, Erasmus was half right when dealing with leaves which, he agreed with Hales, were the lungs of the plant. Hales, he said, had shown that a sunflower lost much more water per unit surface area than did the surface or lungs of animals. The water perspired by those surfaces is hyperoxygenated and, as it escapes from the perspiring vessels, when acted upon by the sunlight, gives out oxygen (as Priestley had observed and Ingen-Housz had confirmed). The oxygen liberated from the perspired water, and added to that of the common atmosphere, presents to the respiratory terminations of the pulmonary arteries on the upper surfaces of the leaves an atmosphere more replete with vital air. The hydrogen liberated when water is decomposed is reabsorbed by the vessels of perspiration. Others believed leaves excreted juices but this seems wrong, said Erasmus, since the vapour exhaled by leaves has no taste (it is indeed pure water) and, as shown by Hales, in moist weather leaves do not exhale or perspire at all. Aware that leaves turn to the light – and anticipating the interests of Charles and Francis – he wrongly ascribed such movements to muscles. He was equally wrong in thinking that the light sought by leaves was necessary for their respiration.

Although willing to try bold and often successful experiments upon his patients, Dr Darwin seldom experimented on plants, which was a pity because whenever he did he made inspired discoveries that challenged accepted dogma. For example, when cutting into the stems of a range of plants and finding that air moved in to replace water, he saw that the vessels remained rigid and did not collapse as in animals. This showed, he said, that vessels are normally filled with liquid, not with air as the currently accepted authorities Nehemiah Grew and Marcello Malpighi thought, and it was further proof that the leaves, not stems, were the lungs of the plant. If the upper surface of the leaf, or lung-equivalent, was specialized for respiration then, thought Erasmus, the lower surface must be specialized for exhalation. He describes how an earlier researcher, Charles Bonnet of Geneva, had put the stalks of leaves into tubes of water and covered either the upper or the lower leaf surface with oil or varnish – Bonnet had found that the upper surfaces exhaled much less than half the quantity of water exhaled by the under surfaces, which shows them to be organs designed for different purposes. (In fact, in most plants the lower epidermis has more stomata per unit area than does the upper epidermis and for this reason transpires more water per unit area). To pursue further the close parallels he saw between the physiology of animals and plants, Erasmus used coloured dyes to trace the veins in leaves. Wanting to see if leaves 'breathed' through the tiny pores (stomata) in their epidermis, he attempted to suffocate several leaves of *Phlomis*, Portugese laurel, and balsams by pouring oil over their surfaces. They died within 24–48 hours and, since insects and wasps which breathe through tiny pores or spiracula in their surfaces also died when subjected to a similar treatment, he concluded there was 'similitude between the lungs of animals and the leaves of vegetables'.[32]

Erasmus proposed correctly that water moves up the stem in specialized vessels, located in a ring just under the surface of the stem, to reach the leaves. However, he accepted Sénebier's incorrect view that this water carried carbon, in the form of carbonic acid, into the leaves (see Table 4.1). In the leaves, water

– shown by Priestley to be composed of hydrogen and oxygen – was either decomposed to oxygen and hydrogen, or it was perspired. Erasmus believed that oxygen entered the leaf through pores and supported respiration. As just seen, he misunderstood why the upper and lower surfaces of the leaf are different and, mistakenly, he thought free hydrogen had an important role in the physiology of the leaf. He did recognize the importance of light to the plant but, ignoring Ingen-Housz, he attributed this to its effect on water rather than the assimilation of carbon from carbon dioxide.

With great insight, he argued that the principal product of 'digestion' in the green tissues of plants was sugar, which is interconvertible with starch:

> The digestive powers of the young vegetable, with the chemical agents of heat and moisture, convert the starch or mucilage of the root or seed into sugar for its own nourishment; ... and thus it appears probable that sugar is the principle nourishment of both animal and plant beings.[33]

It was not until 1845, nearly half a century later, that the German physician Julius Robert Mayer, in a pamphlet entitled *The Organic Motion in its Relation to Metabolism*, recognized energy was conserved during photosynthesis, for the first time expressing the relationship in the simple equation:

$$CO_2 + H_2O + \text{light} = O_2 + \text{organic matter} + \text{chemical energy}$$

The inclusion of water in the equation above post-dates the work of one other man who sits comfortably in the age of Erasmus, even though his greatest work, *Recherches Chimique sur la Vegetation* was not published until 1804, two years after Erasmus's death (see Table 4.1). This man is Nicolas Theodore de Saussure, honorary professor of mineralogy in the University of Geneva. He was not only a colleague of Augustin de Candolle but he had first-hand knowledge of the works of Sénebier who had been an intimate friend of his father, himself a distinguished geologist and mountaineer. Like the Sénebiers, and indeed the De Candolles, the de Saussure family was Protestant; each family had at a different time in the preceding hundred years migrated from Catholic France to the more tolerant Swiss city of Geneva. This background, with its shared shadow of persecution, might have contributed to their tendency to look inward. Thus, despite de Saussure spending some time in England in 1793 and, again, in about 1800, where he made himself familiar with the latest scientific developments,[34] he was reluctant to acknowledge the works of Priestley and Ingen-Housz, and even those of the much earlier Hales.

De Saussure had had an exceptionally good grounding in chemistry and when he began his studies of plants in the mid-1790s was familiar with the latest advances in that subject. He was, like Hales, a great experimentalist. Aware of the need to employ controlled conditions as far as was technically possible, and to duplicate or triplicate measurements, his experiments were designed in a such a way that they yielded quantitative results and incontrovertible conclusions.[35] To give an example, using five different plant species, he enclosed plants in glass vessels containing ordinary air and added known volumes of carbon dioxide. After several hours in the sun, he measured the amount of carbon dioxide taken up and the amount of oxygen released. He measured the increases in total dry weight and carbon that had occurred in the leaves during the experiment. Given the approximations that inevitably arose from the unsophisticated methods of chemical analysis available to him, he was able to conclude that fixed carbon appears in the leaf simultaneously with the disappearance of carbon dioxide from the surrounding air, in which it is replaced by oxygen (see Table 4.1). He had thus demonstrated unequivocally the photosynthetic assimilation of carbon by green plants, something others had been attempting since the time of Hales.

De Saussure argued that since gains in total dry weight were consistently twice the gains in carbon, and since there was no increase in either if carbon dioxide were excluded from the atmosphere, photosynthesis must involve an additional process, the fixation of water (what is today well recognized and called the 'splitting' of water), as indicated in the equation on page 34.

He went on to demonstrate that leafy shoots with their stalks in distilled water assimilated large amounts of carbon from the atmosphere, in doing so finally extinguishing Sénebier's idea that the plant's main source of carbon dioxide is the water it absorbs through its roots, an idea which had misled several of Sénebier's successors.

The few remaining uncertainties associated with de Saussure's investigations were finally resolved in 1860 by an equally talented experimentalist, Jean-Baptiste Boussingault.[36] With improved chemical methods of analysis and in a long series of forty-one experiments involving three flasks – one a control with only water, one with plant material and illuminated, one with plant material and darkened – he measured the gas exchanges of plants ranging from algae to celery. He showed that in photosynthesis the ratio of carbon dioxide fixed to oxygen liberated is close to unity (he had thus established the stoichiometry of the basic photosynthetic equation). He proved that the small amount of 'nitrogen' which de Saussure thought was evolved during photosynthesis was an artefact; it was carbon monoxide formed by the reaction of oxygen with the pyrogallate solution de Saussure had used to absorb the gas as it was evolved by the leaves. And Boussingault completed the long awaited, final and unequivocal demonstration that the opposing processes of respiration and photosynthesis occur simultaneously in the light.

*****

The impact of research on the carbon-based processes, photosynthesis and respiration, from Hales to de Saussure, proved to have little immediate practical impact on agriculture. Boussingault may have quantified the reactants and products of photosynthesis but there was much more to be learned before knowledge of the physiology and biochemistry of photosynthesis could possibly feed into any improvement in agriculture; only in the twentieth century did the first benefits result from this area of research (Chapter 11). As the nineteenth century unfolded, it was discoveries about nitrogen, phosphorus, and the other major inorganic elements needed by plants, rather than carbon, that produced the first benefits for agriculture.

'Burn nothing which may nourish vegetables by its slow decomposition beneath the soil, which constitutes their stomachs'. These words come not from a modern eco-friendly writer but from the pen of Erasmus Darwin.[37] His *Phytologia* was all about the application of scientific principles to agriculture and gardening. If in 1800 there were still huge gaps in botanical knowledge, then Erasmus was quite capable of filling them with remarkably accurate speculations. Thus, he proposed, 'nitrogen ... seems much to contribute to the food or sustenance of vegetables ... and is given out by their putrefaction ... forming volatile alkali [ammonia]'. Also, the 'acid of nitre', from which nitrates form, 'probably may contribute much to promote vegetation'.[38] Sir Humphrey Davy, Professor of Chemistry at the Royal Institution, also recognized the importance of nitrogen for crop growth. Perceptively, he wrote, in *Elements of Agricultural Chemistry* (1813), that 'peas and beans in all instances seem well adapted to prepare the ground for wheat ... it seems that the azote [nitrogen] which forms a constituent part of [their] matter is derived from the atmosphere'.[39] Less perceptively, and despite personally knowing Erasmus, Davy overlooked what Erasmus had recognized, the production of nitrates in the soil and their importance for growth of most (non-leguminous) crops.[40]

Today we know that nitrogen is one of seven major elements needed for plant growth (carbon, hydrogen, oxygen, nitrogen, phosphorus, sulphur and potassium, or CHONPSK). Nitrogen is needed because it is an essential constituent of amino acids, the building blocks of proteins, and of the bases from which DNA and other nucleic acids are built. It is absorbed by plant roots in the form of either nitrate (NO_3^-) or ammonium (NH_4^+) ions, though, in legumes and a few other plants such as *Alder*, bacteria living in root nodules can fix nitrogen directly from the soil atmosphere (Figure 4.1). At the start of the nineteenth century, it was still widely believed that plants obtained their food from humus, the dark layer seen at the top of the soil. Decaying materials gave rise to products that were the most suitable for plant growth. It was difficult for botanists and farmers alike to believe that the insignificant amount of CO_2 in the air, rather than the thick layer of humus coating the soil, could account for plant growth. The benefits of crop rotation, known to the Romans and advocated by

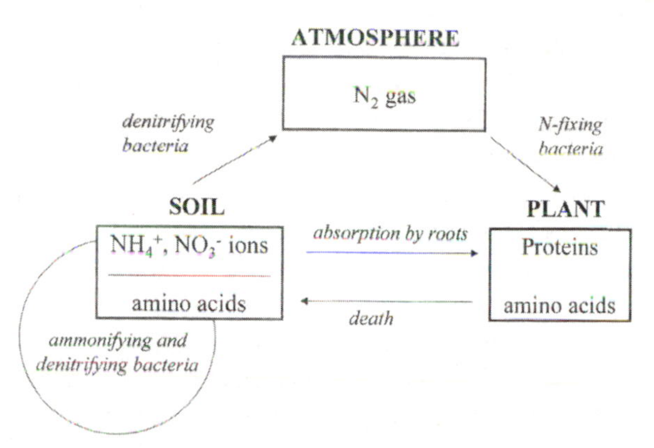

Figure 4.1: The nitrogen cycle. Proteins are broken down to amino acids by either plant or microbial enzymes. In the soil, amino acids are broken down to ammonium (NH_4^+) or nitrate (NO_3^-) ions; either ion can be absorbed by roots, although most plants use nitrate.

such respected authorities as Malpighi, further emphasized the importance of the soil. The immediate practical benefits that resulted from either incorporating a legume into crop rotations, as promoted by the renowned Jethro Tull in 1733 – Tull held the Aristotelian view that small particles of soil could enter the pores of roots – or of manuring, as recommended by Erasmus Darwin, were plain for all to see.

Reviewing the history of botany a quarter of a century later, and with the benefit of hindsight, von Sachs concluded that the 'humus theory' had been one of the last refuges of the vitalists, those who held that the laws of chemistry and physics could not fully explain the functioning of plants and animals, which required an additional ingredient, the 'vital force'.[41] The discrediting of the humus theory in the 1840s was thus one more step preparing the way for the acceptance of Charles Darwin's theory that the evolution of living organisms can be explained completely in terms of the laws of chemistry and physics. Demolition of the humus theory and its replacement by the 'mineral theory' of plant nutrition was achieved largely by Justus von Liebig and Boussingault.[42]

The 'improvement' in agriculture called for by Erasmus and others at the beginning of the nineteenth century was hampered by politics as well as ignorance. In much of western Europe, agriculture and national economies were wrecked by the Napoleonic wars, at the end of which soils were so depleted that there are records of bones being gathered from the battlefields to be crushed and used as fertilizer. The agricultural revolution eventually started by Liebig was both a response to this poverty and a necessary prerequisite for an industrial revolution that was founded upon the mass migration of men and woman from the countryside to the new towns and cities, a migration that left the farming community short of manpower and needing greater productivity from soil and man alike.

Liebig, Professor of Chemistry at Giessen in Germany, enjoyed the highest international reputation. Students flocked to learn in his laboratory and on leaving they carried his view of chemistry around the world. When in 1840 he turned his attention to plant nutrition both chemists and biologists were eager to listen to his opinions expressed forcefully – some would say aggressively – in both lectures and publications. His conclusions were drawn deductively, that is he analysed the ashes of plants grown under different fertilizer regimes (or more often he arranged for other people to provide him with the results of their analyses). He identified what are the major elements necessary for plant growth and, most importantly, he demonstrated humus is not diminished but, in contrast, is increased under healthy vegetation. Once this was established, his calculations proved there remained only one source for the carbon in the plant and that was carbon dioxide in the atmosphere.

Where interpretation of results is concerned, the analysis of ashes has limitations, other than the purely technical ones to do with the sophistication of nineteenth-century techniques. Its problem is that it measures elements, rather than the salts (or, more accurately, the dissolved ions) in which form the elements are taken up. This was nowhere more important than for the element nitrogen where Liebig's methods could not tell him whether it was taken up in the form of ammonium or nitrate ions – though he was (wrongly) convinced it was exclusively the former. He was misled in other ways and, as the years passed, he came more and more to the view that the atmosphere, not the soil, was the main supplier of plant nitrogen. In his 'Law of the Minimum', he proposed a new and what was to prove important principle: plant growth is limited by the element in least supply. Here again, however, he was misled, for his ash analysis did not reveal the relative importance for healthy growth of the different non-limiting, or minor, elements, such as iron or zinc. He was convinced that the key to fertility was to be found in the mineral content of soils and fertilizers, distinguishing mangel-wurzels [*sic*], turnips and corn as 'potash plants', legumes and tobacco as 'lime plants' and cereals as 'silica plants'. Crop rotation was widely accepted by farmers as an important way of maximizing crop yields – in Britain the 'Norfolk rotation' of successive cereal, legume, root and fodder crops was most common – but Liebig recommended most wisely that farmers should adapt their rotations according to local conditions, growing only the sorts of crops that did best for them.

Renewed interest in crop rotation was welcomed in England where the young landowner John Bennet Lawes and his collaborator Joseph Henry Gilbert (a former student of Liebig, who boiled bones in his bedroom to produce the first 'superphosphate' fertilizer) were in 1843 beginning their famous fertilizer experiments at Rothamsted; experiments which, after many years, showed that mineral fertilizers alone were insufficient for the growing plant and that loss of nitrogen from the soil was ultimately the most important factor in its exhaustion. In France, Boussingault too was involved in a long series of crop rotation trials (1836–41) which occupied many hectares of his father-in-law's farm at Bechelbron.[43] The approach of this complex and anxiety-ridden man, who like many of his contemporaries had learned his chemistry at a School of Mining, was, in contrast to Liebig's, very much 'hands-on'. Boussingault reached his conclusions inductively and what he was convinced of by the mid-1850s was that only legume crops increase the nitrogen content of the soil although, if land is constantly well-manured, there is no necessity for a rotation to be followed. Putting down a marker for twentieth-century agriculture, he argued that a monoculture can be successfully followed year upon year if there is sufficient input of nitrogen.

Boussingault's farm experiments were complemented by his own laboratory experiments, though in giving credit for the first laboratory studies of nitrogen assimilation we must turn to de Saussure. He, having established by experiment that nitrogen, unlike carbon, cannot be absorbed by leaves from the air, concluded that nitrogen, like other mineral elements found in the ash of plants, can be assimilated only through roots. Such minerals were necessary to sustain plant growth because, he observed, only stunted development occurred when he grew plants in sand washed with large volumes of distilled water to remove soluble substances. In contrast, near-normal development was observed when he supplied the sand with nutrients in solution, although his analyses revealed that they were taken up selectively, and in less proportion than the water. In small amounts they clearly exerted a large effect upon growth. Like de Saussure, Boussingault too concluded that leaves cannot assimilate nitrogen directly from the air.

What Boussingault established by experiment was that plants will grow vigorously when supplied with nitrogen solely in nitrate-form. Improving on the methods of de Saussure, he grew *Helianthus argophyllus* in sand cultures to which were added, either no nitrogen, or plant ash plus potassium nitrate, or plant ash plus potassium carbonate. After eighty-six days he found that only plants of the second treatment flourished and it was only in those plants that the nitrogen content had increased significantly. Boussingault was still not sure of the relative importance of the two common forms of nitrogen available to plants roots, nitrate-nitrogen and ammonium-nitrogen. Into the 1860s, he continued with his ingenious experiments exploring the absorption of the two forms of nitrogen when supplied in mixture to plants but, sadly, he was finally defeated by the contemporary lack of suitable methods to distinguish reliably the two forms.

Boussingault's outstanding work on nitrogen and crop growth has been oddly overlooked, as for example by Charles Darwin, but it was certainly admired and extensively cited by J. B. Lawes and J. H. Gilbert who, by contrast, received the unstinting praise of Charles and most of the Anglophone world. By 1861, when the Rothamsted workers reported their extensive field trials, together with the supportive laboratory experiments of a young American visitor, Evan Pugh, the conclusion was clear.[44] As a table from their paper in the *Philosophical Transactions of the Royal Society* shows, legume crops, such as bean, were unequivocally different. At harvest, they contained much more nitrogen than other crops grown under similar circumstances: legumes must, as De Saussure, Boussingault, and Lawes and Gilbert themselves, had long argued, be able to tap a source of nitrogen not available to all other crops.

Table 4.2: Results of Lawes, Gilbert and Pugh's Experiments into the Nitrogen Yields of Different Crops at Harvest.

Crop	Dates of experiments	Number of years	Average yield of nitrogen per acre, without manure, lbs
Wheat	1844–59, inclusive	16	24.4
Barley	1852–9, inclusive	8	24.7
Meadow hay	1856–9, inclusive	4	39.4
Beans	1847–58, inclusive	12	47.8

Only a few more years were to pass before that source was demonstrated by experiment. Its revelation had to await the birth of microbiology, but soon, between 1886 and 1888, a series of papers was published by Hermann Hellriegel and Hermann Willfarth showing that nitrogen was directly fixed from the atmosphere into the root nodules of broad bean (*Vicia faba*), and other leguminous plants[45] –nodules first observed by Malpighi. To be more precise, nitrogen was first fixed by *Rhizobium* bacteria living in the nodules before being released into the surrounding root tissues, but the point was made. The peculiarity of legumes was at last explained.

De Saussure, Boussingault and the Rothamsted workers established irrefutably the importance of nitrogen for plant growth. Their scientific studies provided a rationale for practical improvements in agriculture, which would have delighted Erasmus Darwin. And, together, they provided a context in which, when Charles Darwin began his studies of carnivorous plants (Chapter 6), he could better understand their particular nutritional need for nitrogen.

*****

As chemistry complemented botany and the parts played by mineral nutrients, carbon and water in the growth of plants were slowly revealed, botany began to escape from the weight of general ignorance and of vitalism in particular (though see Chapter 5). Complete explanations for biological phenomena were increasingly found in the laws of physics and chemistry. An intellectual climate was established in which new ideas and hypotheses were routinely tested against evidence. Mindsets and habits were prepared for one of the most shocking new theories ever to challenge man's intellect – evolution by natural selection.

5 CHARLES DARWIN'S EVOLUTIONARY PERIOD

That so great a man as Charles Darwin could be so well liked by his contemporaries was due in no small part to his natural modesty. On scientific matters, as in everyday life, he was approachable, naturally tending throughout his life to assume the part of the eager student rather than the overbearing teacher. His modesty was nowhere more marked than in his attitude to botany. On being elected to the illustrious French Institute in 1878, he wrote to his old friend Asa Gray, 'It is rather a good joke that I should be elected in the Botanical Section, as the extent of my knowledge is little more than that a daisy is a compositous plant, and a pea a leguminous one'.[1] To the very end of his life he protested that he was not a botanist; or he would have people believe that he was, 'one of those botanists who do not know one plant from another'. Such claims fooled few of his contemporaries. Charles was not as botanically naïve as he would have had others believe.

From earliest childhood, he had learned and read extensively about plants and the botanical discoveries of explorers, such as Joseph Banks and Alexander von Humbolt – he particularly admired the latter's books and re-read them several times.[2] He did not study botany in any formal sense while at Cambridge, but he had forged an exceptionally close relationship with the Professor of Botany. After leaving the university, he had studied the plants of South America during the voyage of the *Beagle* and when he settled with Emma at Down House he immediately began to use his garden and greenhouse to make extensive observations on botanical phenomena as diverse as mechanisms of pollination and the dynamics of seedling establishment. As this and the next chapter will demonstrate, Charles's comments were, for once in his life, intended to mislead. By the standards of the day he had received a sound education in classical, or descriptive, botany. Moreover, his education was continually updated thanks to his friendship with Hooker, Gray, and others who more readily admitted their botanical expertise.

As a young man, he may have known *relatively* little, so it is possible to understand his protest on returning from the voyage of the *Beagle* in 1836:

> I felt very foolish when Dr. Don remarked on the beautiful appearance of some plant with an astounding long name, and asked me about its habitation. Some one else seemed quite surprised that I knew nothing about Carex from I do not know where. I was at last forced to plead most entire innocence, and that I knew no more about plants which I had collected than the man in the moon.[3]

All young men, however, by dint of their few years, know *relatively* little. Knowledge accumulates with age. Why then did he continue in later life to play down his botanical expertise? In the background there was, of course, his modesty. He was delighted whenever one of his botanical works was praised by an 'expert' botanist, particularly if that expert were Hooker or Gray. A simple explanation for his continued self-denigration is that, as so often in his life, he was extremely nervous of criticism. In botanical matters he sought to justify his possible errors (and there were very, very few of them) by pleading his ignorance of the subject. There is one possible exception where, either consciously or unconsciously, his motivation was different and he used pleas of ignorance to gain practical advantage. As will be seen, his aim was to enlist the help of his old professor but, in the event, his plan failed.

*****

Charles's earliest interest in plants, indeed in all aspects of the natural world, was kindled by his father, Robert, whose interest had, in turn, been ignited by the botanical rarities collected by his own father, Erasmus. At The Mount, Robert's home in Shrewsbury, the collection of garden plants had been enthusiastically extended, while a hothouse had been added in order to accommodate tender tropical plants, Robert's special interest. Charles's earliest lessons were strictly practical ones. Robert taught his boy not just the names of plants but, more significantly, he enlisted his son's help in the maintenance of an extensive 'garden book' in which were recorded the various plantings they made.[4] Further notes were made about the development of the garden. Focussing on such details as the number of flowers on the Peonies, Robert introduced his son to the habits of observation, counting and cataloguing. No amount of skilled teaching can ignite enthusiasm in a child if he or she has no natural interest in a subject but, in Charles's case, there was no lack of interest in, or love of, plants. Francis affectionately remembered his father's dedication as:

> a kind of gratitude to the flower itself, and a personal love for its delicate form & colour. I seem to remember him gently touching a flower he delighted in.[5]

The same strong fascination with plants and animals burned undiminished in the minds of generations of Darwins. Like all collectors, they enjoyed possession and organization, in their case not just of plants but of information about plants and their world. Charles recalled:

> From my earliest days I had the strongest desire to collect objects of natural history, and this was certainly innate or spontaneous, being probably inherited from my grandfather.[6]

However, Charles was more than a collector of objects and facts for, like his grandfather, his mind was open to the possibility of change or improvement:

> Apparently I was interested at this early age [first school] in the variability of plants! I told another little boy ... that I could produce variously coloured polyanthuses and primroses by watering them with coloured fluids.[7]

If lessons in the garden taught Charles from an early age how to cultivate plants, and even to quantify their growth, it was from the books he read – mainly during the holidays from boarding at Shrewsbury School, where he went from the age of nine – that his informal education was completed. By searching the extensive library at The Mount he could find out most of what there was to be known about the anatomy, classification and geography of plants in the first quarter of the nineteenth century. Extraordinarily, the growing boy had the rare privilege of knowing that some of the most important books in the botanical collection had been written by members of his own family; by his grandfather, Erasmus, and by his great uncle, Robert Waring senior. Few boys can have been so fortunate.

Equally luckily for the young man, if not for botany, there was little risk that Charles's botany would be out of date by the time he arrived in Cambridge in 1828. Since the publication of *Phytologia* in 1800, the sum of knowledge about plants had grown very slowly in the absence of innovative thinkers and investigators. In the intervening twenty-eight years the dominant activities had been the recycling and occasional reinterpretation of old knowledge.

At Cambridge, where he went to study theology (see Chapter 2), Charles found endless distractions. To the displeasure of his father, he spent most of his time in the local countryside riding or shooting, often in the company of his cousin William Darwin Fox. What little time he had left he devoted to his current passion, collecting beetles. His painful experience of attending lectures in Edinburgh had convinced him that he should avoid a similar experience in Cambridge, although during his second year in Cambridge he relented, attending lectures given by the Professor of Botany, John Stevens Henslow.

> I attended ... Henslow's lectures on botany, and liked them for their extreme clearness, and the admirable illustrations ...[8]

He was persuaded to do so by the personal recommendation of his brother, Ras, one of the few people whose advice young Charles would take. Ras, who was more open minded than his brother about the value of lectures, had himself experienced and liked Henslow's lectures. The other factor which probably persuaded Charles to make an exception where Henslow's lectures were concerned was that he already knew the professor socially by virtue of his having attended soirées organized by the professor and his wife, Harriet.[9]

Naturally, because the botany course was heavily orientated towards the needs of medical students, there was an emphasis upon recognizing plants and their medicinal properties. However, Henslow was clearly an exceptional teacher. A devout Christian, who later relinquished academic life in order to work as a parish priest, he delivered daily lectures through the spring term on topics ranging from the theory of classification to plant function, the latter including subjects not usually taught to undergraduates, such as the function of sap, or the process of seed germination.

Where his approach to the classification of plants was concerned, it has been appreciated only recently how Henslow's *modus operandi* may have had an important influence upon Charles. Like many botanists Henslow collected wild plants, which he dried and pressed on sheets of herbarium paper, in which form they could be stored indefinitely and studied at leisure. What Henslow did quite uniquely, however, was to display on each sheet a 'collation', that is several plants arranged to show the variation in form within that species. As illustrated by his *A Catalogue of British Plants* (1839), his natural tendency was to 'lump' together as varieties of one species those differing forms which others might 'split' into different species. Where Henslow saw variation within fixed species created by his God, Charles saw variation that – he would one day realize – was the raw material of evolution and the origin of new species.[10]

In the year Charles entered Cambridge, 1828, Henslow had printed a sixteen-page *Syllabus* for his students. It included a reading list. On the list were two books by James Edward Smith, *Introduction to Physiological and Systematic Botany* (first published in 1807, then in its sixth edition) and *English Flora*, and also two books by Augustin-Pyramus de Candolle, both in French. Smith was the highly respected founder of the Linnaean Society. When writing to his strengths – descriptive botany and systematics, strictly in the Linnean tradition – he was excellent, but when writing about plant function he was at his weakest, having to lean heavily on Stephen Hales and Erasmus Darwin for his information, as the abundant references to those men show and which, to his credit, he openly and readily admitted. De Candolle was also a systematist who, along with the De Jussieaus in France (as noted in Chapter 3) and William Hooker and Robert Brown in England, was gradually replacing the rigid sexual system of Linnaeus with those more 'natural' systems of classification for which Erasmus had argued. Such systems were based on morphological features and the positional relations of organs.

By 1833 Henslow had replaced his *Syllabus* with a *Sketch of a Course of Lectures*. Fifteen lectures, more than half the course, were devoted to what was labelled 'plant physiology', a very broad attempt to relate anatomy and morphology to function. To the list of recommended reading was added de Candolle's *Physiologie Végétale* (1832), while Smith's books had been replaced by John Lindley's *Introduction to the Natural System of Botany* (1830) and *Introduction to Botany* (1832). Lindley's writings promoted the new 'natural' systems of classification but beyond that, like Smith, he displayed little originality being too firmly grounded in the stale traditions of listing and naming. When inaugurated as the first Professor of Botany in the University of London, in 1829, the most exciting thing he could find to say was:

> Let the vegetable world be studied in all its forms and bearings, and it will be found that certain plants agree with each other in their anatomical condition, in the venation of their leaves, in the structure of their flowers, the position of their stamens, in the degree of development of their organs of reproduction, in the internal structure of those organs, in their mode of germination, and finally in their chemical and medicinal properties.[11]

He had missed Erasmus's exciting message that a study of function might lead to the improvement of plants. Lindley betrayed the boldness of his employers in establishing the new Chair for he was meekly content that botany should remain shackled to medicine.[12]

De Candolle is significant for his strong influence on Henslow's botany. His books were on both reading lists and, it seems reasonable to infer, his views must therefore have been respected and repeated to Charles by Henslow. (Charles certainly owned a copy of de Candolle's *Theorie Elementaire de la Botanique* – as well as copies of several of Smith's books – *before* he embarked on the voyage of the *Beagle*).[13] Not surprisingly, when Henslow published his *Principles of Descriptive and Physiological Botany* (1835) its heavy reliance on de Candolle was obvious.

Since de Candolle's writings were clearly familiar to Charles at second hand, if not at first hand, they are worth exploring in more detail before returning to Henslow. They reveal not just what was known but, just as importantly, what was not and how the gap was filled in a way supportive to evolutionary theory. De Candolle's strength was, as mentioned before, taxonomy and systematics. When describing function he was uncritical and relied on a narrow basis of information coming almost exclusively from French sources. He especially favoured those known to him in Geneva (it was Sénebier who had urged the young de Candolle to apply himself to vegetable physiology). Hence, he referred to the works of Bonnet, Sénebier, Dutrochet and de Saussure, but mentioned Hales only *en passant*. In spite of having visited London in 1816,[14] when among others he met Robert Brown and James Smith, and in spite of clearly having some grasp of English, the only English language author to receive much attention from de Candolle was Andrew Knight (of whom there will be more later) but this is probably because his fellow Frenchman, Henri Dutrochet, repeated Knight's most famous experiments on tropisms (see Chapter 7).[15]

De Candolle's great work, *Physiologie Végétale*, was innovative in several ways. Firstly, physiology had its own separate part in the book, although little detail was given of experimental technique and there were few diagrams. Secondly, de Candolle wrote about crops as well as wild plants and, just as unusually for the times, he presented in some chapters quantitative data originating from the crops or wild plants he had studied. Unfortunately, the data was selected in a way that would have been unhelpful to his contemporaries. Thus, although he detailed the water content of various fruits before and at maturity, and he tabulated the yield of cereals, beans and other crops on contrasting soils (he was an advocate of the beneficial effects of crop rotation), his results represented what had happened under particular, often peculiar, sets of conditions, making it difficult for general lessons to be drawn. Similarly, where natural vegetation was concerned, few fundamental truths could be extracted because he reported observations made on plants growing at specific, and often unusual, sites, such as the Cevennes or the Isle of Guernsey, whose conditions were unlikely to be replicated elsewhere.

To counterbalance these positives, there were plenty of negatives. *Physiologie Végétale* was lengthy, discursive, and poorly referenced. And there were errors of fact; errors which sadly were repeated by later authors. For instance, de Candolle believed in the existence of spongioles, writing the 'spongioles' of roots, being actively contractile and aided by the capillarity and hygroscopic qualities of their tissue, suck in the water that surrounds them together with the saline organic or gaseous substances with which it is laden. He proposed also that the water that has been sucked in moves through the wood and intercellular spaces to the leaves. Water exits via the stomata, he correctly noted, but then added in error that it leaves behind carbonic acid which is 'decomposed' in the light. The leaf also contains, he thought, whatever water is derived from respiration and a little that is absorbed directly. A nutrient sap descends from leaves to roots during the night. There is no circulation like that in animals but there is an alternate ascent and descent. Both phenomena depend on the contractile properties of young cells, the power for which is supplied by the vital force. The idea of spongioles was accepted and repeated by Henslow, among others, although, in a rare disagreement with de Candolle, Henslow did reject the idea that there is dilation and contraction of intercellular passages.

In his *Principles of Descriptive and Physiological Botany*, Henslow reproduced some serious errors from de Candolle. Thus, just as de Candolle sometimes lapsed into explanations based on the intervention of a 'vital force' or 'vital principle', so did Henslow. It is difficult for the modern mind to grasp what was meant by the 'vital principle', but it is basically what is left when chemical, physical and mechanical causes are taken away. Henslow recruited the vital force in other areas when other explanations failed him, as in the example below (in this instance compounding the misinformation by recommending experiments on *dead* tissues):

> In the complex phenomena which vegetation furnishes, it is very difficult to separate so much of each result as may be strictly ascribed to the operation of the vital principle, from such as may be due to the action of purely physical causes, the chemical effects of affinity, and the mere mechanical properties of the tissue. The most obvious means which we can employ, for ascertaining the precise properties of the tissue, is to perform experiments upon it in the dead vegetable, and as nearly as possible before any chemical change may have taken place in it.[16]

Henslow's book was much shorter than de Candolle's, but was similarly poorly illustrated and referenced – even Charles was frustrated enough in later life to point out this deficiency in his old professor's book.

Thankfully for the undergraduate Charles, Henslow attached much importance to field botany, which suited the young man's approach to learning: Charles recalled:

> Henslow used to take his pupils, including several of the older members of the University, on field excursions, on foot or in coaches to distant places, or in a barge down the river, and lectured on the rarer plants or animals which we observed. These excursions were delightful.[17]

Much has been written about the relationship between Henslow and Darwin, a relationship between professor and student as unusual then as it would be now, but one which steadily built into a lifetime's friendship. Charles was regularly invited to the soirées at which there was tea and conversation between a group of twelve to fifteen carefully selected guests. By the end of his undergraduate career in 1831 Charles knew the family closely, often joining them at their meal table. He was such an obvious favourite that other dons nicknamed Charles, 'the man who walks with Henslow'.[18]

Young Charles hung upon the older man's lips and words, even competing for Henslow's attention with another exceptionally able student and potential favourite, Charles Babington. Darwin could justifiably have been thought by his fellows to be something of a lickspittle. There is a story that on a botanical excursion to Bottisham Fen the students were provided with poles to help them jump across the ditches. In order to secure a particularly good specimen of an aquatic insectivorous plant, *Utricularia*, he attempted a pole jump only for the pole to become firmly stuck in the mud in the middle of the ditch. Charles coolly slid down into the muddy waters, collected the specimen and brought it, 'all besmirched', to the amused professor.[19]

In their superb biography of Henslow, Walters and Stow called him simply, 'Darwin's Mentor': his teaching and advice extended far beyond the strict confines of botany. It was Henslow who recommended Charles for the post of naturalist on the voyage of the *Beagle* to South America (Henslow's brother-in-law, Leonard Jenyns, having already declined to take the post). And it was Henslow who recommended Charles to take with him on the *Beagle* a copy of Lyell's 'new book', volume one of *Principles of Geology* (1830), warning it is 'altogether wild as far as theory goes'.[20] This theory was the 'uniformitarian' view that the forces of nature assumed to have acted over geological time were the same as those observable today. This was in contrast to the 'catastrophist' view, which in its extreme form assumed the biblical creation and flood. The book changed 'the whole tone' of Charles's mind and, ultimately, was a major factor contributing to his thoughts on organic evolution.

Whenever opportunities arose during the voyage, Charles and Henslow corresponded about the latest news in Cambridge, mutual friends, and what plant material Darwin should collect or send by packet boat back to England. As soon as the voyage was over in October 1836, Charles sent ahead a large consignment of plants to Henslow. Although Charles could bask for the moment in the warm glow of his own outstanding achievements there was hanging over him the serious problem of working out how best to record and archive the huge collection of specimens. He sought the help of Henslow.

To this end he was happy to portray himself as the botanical ingénue, as shown by the above cited claim that 'he knew no more about [the] plants ... than the man in the moon'. This was just not true. For one thing, he must have been able to recognize those plants that Henslow told him to collect from South America. For another, from childhood, let alone from his zoological experiences, the importance of recording the details of the habitats from which specimens are collected had been drilled into him. It was ingrained so that although some *Beagle* specimens carry the simplest comment, 'plant', most have much more detailed notes, suggesting they were collected to a well thought-out plan. Certainly, in collecting plants from the Galapagos Islands each was carefully labelled, each specimen with the name of the island and date of collection.[21] His notes, or rather those written by his servant and amanuensis, Syms Covington, sometimes ran to a page or more, as in the description of a wild potato from Chile, *Solanum tuberosum* var. *guaytecarum*, or of the habit of the 'Dwarf Beech' of Tierra del Fuego, *Nothofagus antarctica*.[22]

Charles's strategy was to persuade the overworked Henslow to help him identify, preserve and archive the 600 or more plants he had collected from South America. To maintain the pressure, he even told Henslow how much the mutually admired von Humboldt wanted him to undertake the task!

Charles's strategy was ill conceived, for Henslow had struggled to keep up with the plants that Charles had despatched to him while he was away, so he was quite unable to cope when the bulk of the plant collection was delivered to him. Clearly incapable of devoting the time that was merited by such a large and important collection, he procrastinated.[23] Progress stalled for several years until Henslow finally admitted to Charles that at least part of the collection should be passed to someone who could give more time to it. He recommended a young botanist with a fast growing reputation, Joseph Dalton Hooker (who, incidentally, married Henslow's daughter, Frances, in 1851). The recommendation could not have been more appropriate. Hooker had in 1843 just returned to the Royal Botanic Gardens at Kew after four years spent in the Antarctic as the surgeon, but *de facto* naturalist/botanist, with James Clark Ross's expedition. Hooker was looking for a new challenge and his admiration for Charles pre-dated the polar expedition. He jumped at the opportunity of working on Darwin's collection. At last, serious work began and there began also a friendship between Charles and Hooker, based on mutual admiration, that was to last a lifetime. The two men shared triumphs and tribulations and, it is said,[24] Charles's feelings towards Hooker became more like those of a brother. In the Introduction to *On the Origin of Species*, it was Hooker who was singled out to receive Charles's gratitude and admiration:

> want of space prevents my having the satisfaction of acknowledging the generous assistance which I received from very many naturalists, some of them personally unknown to me. I cannot, however, let this opportunity pass without expressing my deep obligations to Dr Hooker, who for the last fifteen years has aided me in every possible way by his large stores of knowledge and his excellent judgment.[25]

Charles had fleetingly met Joseph Hooker, and also the American Asa Gray, that other mainstay of his botanical work, in 1839 during Gray's first visit to Britain. Gray's letter of introduction had been to William Hooker – Joseph's father and at the time Regius Professor of Botany in Glasgow – but William soon recognized an affinity between the two young men and left Joseph the testing task of introducing Gray to many of 'the great and the good' of the day. When visiting the distinguished anatomist, Richard Owen, at the Royal College of Surgeons, they met by chance the young explorer Charles Darwin who happened also to be visiting on the same day, although, compared with Owen, Darwin made little immediate impression on Gray.[26]

Joseph Hooker was a plant collector and taxonomist who through his travels to India, and New Zealand, and later through his Directorship of the Royal Botanic Gardens, Kew – where he succeeded his father – made connections with and between botanists and Botanic Gardens throughout the world. Asa Gray was a taxonomist and also a plant geographer. At the time, American botany lagged somewhat behind European botany, so, of necessity, Gray regularly corresponded with, and took every opportunity to visit, Europe's leading botanists. Amongst his friends was the Swiss plant geographer Alphonse de Candolle, son of Augustin for whom both he and Charles had had such great regard. (In later years, Alphonse himself frequently corresponded with Charles and in 1882 he even visited the old man at Down.)

The expertise of Hooker and Gray both reflected and to an extent limited Charles's botanical interests during the years leading up to publication of the first edition of *On the Origin of Species*. Writing to Asa Gray in 1886, many years after their friendship had been forged, the ageing Hooker said,

> I ... have thrown aside all idea of making headway with – any desire to keep up with even – heads of Chemico-botany, and Micro-phytology. I may content myself with a casual grin at young men calling themselves botanists, who know nothing of plants, but the 'innards' of a score or so.[27]

Gray's reaction is not recorded but he was almost certainly sympathetic. He was no physiologist; his own book, *Elements of Botany*, had presented what was then (1836) the old, de Candollian, view of what constituted physiology.

Francis Darwin was in no doubt about the value of Hooker's advice to Charles:

> it was not merely information, guidance, explanation, that he received, but an inspiriting companionship, a fresh and vigorous influence giving continuous refreshment to the solitary worker. ... The following list of subjects, taken at random from my father's letters to Sir Joseph, give an idea of the subjects discussed during the evolutionary period:- The dispersal of seeds, continental extension, geographical barriers, the arctic Flora, alpine plants, large genera varying, coal plants, island Floras, aberrant genera, direct action of conditions, rarity and extinction, sterility, graft hybrids.[28]

Many of these letters between Darwin and Hooker also involved Gray for the men kept up a regular three-way exchange of ideas and information, the serious science being liberally seasoned with friendly banter and personal news. However, Charles was quite masterly in the way that, while acknowledging his friends were overworked, he would, often in the same letter, pepper them quite mercilessly with questions, or appeals for materials. Thus, on 26 June 1863 he wrote to Gray:

> Although it is one of my pleasures to write to you & a very great pleasure to receive a letter from you; I earnestly beg you *never* to write to me when so busy; if I did not hear for six months or twelve months, I should understand the cause.

And a few lines later, in an oblique request for immediate help:

> The seeds of Sicyos did not germinate: & only one plant has come up of Echinocystis [Gray had supplied the seeds].[29]

On 25 August 1863 he wrote to Hooker:

> Whenever you write please tell me the **Vol. Title** of Journal, **Year**, & page of your paper on the 'Climate &c' of the Himalaya. And I ask you, whether you ought not to be crucified alive for sending out a valuable pamphlet with no means of giving a reference?

Although a few lines later, he tempered his request:

> Do not write until you have something approaching to leisure.[30]

Asa Gray was instrumental in converting Charles's passive interest in climbing plants into a major, active research programme, as will be seen later, but without Hooker, thought Francis, his father would have been unable to bring about the revolution in botanical geography that he did.[31]

Close working ties between the men were underpinned by family friendships. The Hookers and the Darwins visited each other's homes, with the warm feelings between Charles and Joseph being matched by those between their wives. When Charles's growing sons, Francis and George, made their first visit to the US in 1871, Charles arranged for them to visit the Grays, thus securing potential help and support for the two young travellers, while at the same time strengthening ties between the two families.

One other friendship from this period between the *Beagle*'s voyage and publication of *On the Origin of Species* is particularly noteworthy, and that was between Charles and the elderly Robert Brown, possibly the most eminent British botanist of his day. For over forty years Brown had sat spider-like at the centre of a botanical web.[32] He had been librarian at Joseph Banks's home in Soho Square, London (where John Lindley had been his assistant), and from 1827 was the first keeper of the Botanical Department of the British Museum. He was above all a single-minded collector who, in Charles's words, was miserly about his dried plants and fiercely protective of his acquisitions. However, he was also an exceptionally talented microscopist and, while investigating the fertilization process in orchids, he became in 1833 the first man to recognize those two fundamentals of all cells, the nucleus and that universal and constant motion of the cytoplasm which has become known as 'Brownian Motion'. Before sailing on the *Beagle*, Charles had sought Brown's advice on the best sort of microscope to take on the voyage. On his return, and prior to his marriage to Emma, Charles saw a good deal of Brown. Living in London,

> I used to go and sit with him almost every Sunday morning. He seemed to me to be chiefly remarkable for the minuteness of his observations and their perfect accuracy.[33]

Charles could have had no better teacher.

> His knowledge was extraordinarily great, and much died with him, owing to his excessive fear of making a mistake. He poured out his knowledge to me in the most unreserved manner, yet was strangely jealous on some points.[34]

No doubt, the subject of fertilization came up. By virtue of the hours he had spent under Henslow's tutelage in Cambridge observing the fertilization process at first-hand, Charles would have been able to appreciate the depth and extent of the old man's knowledge. It was a subject to which he was to return and which was to occupy much of his time in the years after *On the Origin of Species*.

*****

Publication of *On the Origin of Species* was the defining moment in Charles's life as the never-ending flood of commentaries, criticisms, articles and books written since 1859 adequately testify. His reputation and personal life were changed forever, but the event also marked in strictly scientific terms a beginning as much as an end. What Francis, his biographer, clearly recognized, and what has only recently been confirmed by others,[35] is that around 1860 a fundamental shift of emphasis occurred in Charles's work. The balance of his attention swung from animals to plants and he moved from what Francis called his 'Evolutionary Period' to his 'Physiological Period'. The transition was neither sudden nor absolute – his book, *Orchids* (1862), belongs firmly with his evolutionary period, as do his papers on the forms of flowers delivered to the Linnean Society between 1861 and 1868 – and in many ways the second period complemented the first. It has been argued that *On the Origin of Species* can seen as an heuristic prologue serving to guide Charles's investigations over the next two decades.[36] What is beyond dispute is that, from roughly age fifty-one onwards, Charles's became increasingly interested in the physiology of plants (the topic of the next chapter).

In his Evolutionary Period, Charles relied heavily upon on the advice of expert botanists such as Gray, Hooker and a German contact, Fritz Müller.[37] *On the Origin of Species* is peppered with references to the works and ideas of these men and to the two de Candolle's, Augustin and his son Alphonse. This small group of travellers gave to Charles eyes in the Himalayas and Antarctica, in New Zealand and North America, distant lands he would never see at first hand. They furnished him with examples from exotic and extreme environments that he used to expand and illustrate his arguments about migration and extinction, about selection and hybridization.

Charles was finally pushed into publication by his receipt of Alfred Russell Wallace's letter which, in spite of its shortness, encapsulated Charles's theories of evolution and natural selection. He realized he had to move quickly. He saw *On the Origin of Species* as a stopgap, a short publication that would indelibly establish his own precedence. At a later date, he thought, he would complete his long planned 'big book' incorporating the wealth of data that he had accumulated over many years and for which there was no room in this original volume. Charles never did write his big book but he did revise and extend *On the Origin of Species* through six editions, the last appearing in 1876. In parallel, he wrote a series of articles and short books based not so much on what he had accumulated over decades but, rather, on new knowledge that he acquired in the years after 1859. This he obtained either by further interrogation of his correspondents or through original researches which he undertook himself. The publications range in subject from *Variation of Animal and Plants Under Domestication* (1868) and *Effects of Cross and Self-fertilization* (1876) – which are both related to his general interest in hybridism, as is his chapter on the same subject in *On the Origin of Species* – to *Orchids* (1862) and *Insectivorous Plants* (1875), which at first sight have no obvious connection with his previous interests. What connected them was his motivation. The common purpose of his various works was to confirm and embellish evidence and arguments contained within *On the Origin of Species*. He wanted to convince his doubters and, although he may not have known it himself, to bolster his own occasionally flagging convictions.

Argument based on evidence was Charles's greatest strength, possibly his defining characteristic as a scientist. Excluding those chapters in the first edition of *On the Origin of Species* dealing with geology or with animal instinct, the number of examples drawn from the plant kingdom was similar to the number drawn from the animal kingdom. The only detectable difference is that where animal examples were cited the text tended to be longer, giving greater detail and often drawing on Charles's own observations. Where writing in more general terms he was scrupulously even-handed, time and again using the couplets 'animals and plants' do this, or 'plants and animals' do that. As *On the Origin of Species* went through successive editions, so sections of it were revised and expanded, with specific examples and cases being introduced from his parallel publications. Most importantly, the botanical evidence was reinforced.

*****

Charles's books about *Orchids, Insectivorous Plants* and *Climbing Plants* are outstanding works of scholarship covering areas of botany that had until then received very little attention. Their merits have withstood the test of time. His books on *Variation* and *Cross and Self Fertilization* are a different matter for, while his basic observations were correct, the explanatory framework into which he wove them became outdated with the advent of Mendelian genetics at the end of the nineteenth century.

During the voyage of the *Beagle* Charles had collected cacti, orchids and other exotic blooms and, on his return, the form and function of flowers increasingly occupied his thoughts. It was probably his longest standing botanical interest, his arguments on the subject proving central to many of his papers and books (see Table 6.1). In most flowering plants, male and female organs occur in the same flower so, unsurprisingly, most botanists had accepted the obvious conclusion that self-fertilization was the norm. Equally unsurprisingly, Charles did not accept the conventional wisdom, believing instead that it was normal for pollen to be transferred *between* flowers, thereby effecting cross-fertilization and generating variability among offspring, variability that was the raw material for natural selection. He was bolstered in this belief by the earlier writings of the well respected botanist Andrew Knight whose horticultural studies of fruit trees had led him to argue that cross fertilization, achieved with the aid of insects, was the norm. In 1799, Knight had written:

> In promoting this sexual intercourse between neighboring plants of the same species, nature appears to me to have an important purpose in view.[38]

Using *Linaria vulgaris* (yellow toadflax), *Dianthus caryophyllus* (carnation), and several other plant species, Charles conducted innumerable tests to find the effects of in-breeding on normally out-breeding species. His 1864 results with *Linaria* were most striking. Following his usual method he had covered some plants with gauze netting to prevent their (cross) fertilization by insects, while flowers of other plants remained uncovered and open to the normal pollinators. The five best pods from the covered plants produced together a total of only 118 seeds, while just one pod taken from uncovered plants produced 166 seeds. When the seeds from the two sources were sown under identical conditions in his glasshouse, the seedlings from the self-fertilized plants were shorter and altogether less robust than those from cross-fertilized plants.

If outbreeding was so beneficial, might there be adaptations to favour it, he asked. He observed that among the mechanisms flowering plants use to maximize out-breeding is dichogamy, the maturation of the male and female organs of the same flower at different times. Another mechanism he noted is where only male, or only female, organs mature within flowers on one plant so, effectively, plants are either male or female. (About 4 per cent of flowering plants are dioecious, i.e. they separate the sexes in this way, while, in contrast, sexual separation is the norm in animal species).

The mechanism of ensuring cross breeding that captured his interest more than any other was dimorphism. When Charles was a student, Henslow had introduced him to the mysteries of *Primula* spp., which had perplexed botanists since the time of John Ray, nearly two centuries earlier.[39] Probably, memories of his student days were revived by what he saw in the fields and hedgerows around Downe for, in spring, they were filled with bright yellow carpets of primroses (*Primula vulgaris*) and cowslips (*Primula veris*). There was uncertainty as to whether these two – and also a third plant, the oxslip – were one species, as Henslow believed, or were separate. There was a widespread belief that primrose seeds could give rise not just to primroses but also to cowslips. However, when transplanted to Charles's greenhouse they did nothing of the sort, the two plants bred true; the primrose and cowslip were separate species. Now, in drawing half-sections of flowers of cowslip, Henslow had clearly depicted two different forms: one having a short style and long stamens, and called the 'thrum' morph, the other having a long style and short stamens, and called the 'pin' morph. Charles had probably seen Henslow's drawings but, whether he had or whether such observations were new to him, he now made an intellectual leap forward by recognizing the functional significance of dimorphism; it promoted cross breeding. Pollen from flowers with long stamens would be transferred by insects to flowers with long styles; pollen from flowers with short stamens would be transferred on different parts of the same insect, or by different insects, to flowers with short styles.

Primula, and similarly *Linum grandiflorum* (garden flax) which he found also showed dimorphism, or 'heterostyly', were relatively easy to understand. He was, he told Gray, driven 'almost stark raving mad',[40] when he moved on to explore a trimorphic species, *Lythrum salicaria* (purple loosestrife). The three forms differed in the lengths of their pistils and stamens, of which there were two sorts in each flower, and in the size and colour of their stamens. The study of *Lythrum* had all the appearances of a self-imposed nightmare but he declared that nothing in his life had interested him more and, with the advice of Gray and drawing on the mathematical expertise of his son George, not to mention the collecting skills of his Wedgwood nieces, he was finally able to find the solution. This he presented to the Linnean Society in 1864. For full fertility, he had deduced, the stigma of one form should be fertilized by pollen taken from the stamens of corresponding height in another form; any other (illegitimate) combination resulted in fewer than normal seed which, in their turn, produced sickly, sterile plants.

Charles wrote, 'Botanists, in speaking of fertilization of various flowers, often refer to the wind or to insects as if the alternatives were indifferent. This view, according to my experience, is entirely erroneous'.[41] His message to botanists was emphatic: each plant had evolved to be pollinated by just one of the two alternatives, wind or insects. The evolutionary pressures were different; those using wind produced pollen in large amounts and had, relatively, simple flowers designed to trap pollen from moving air; those using insects produced pollen in smaller amounts and had typically developed larger showy flowers incorporating colours, scent, and sometimes nectar to attract flying or crawling insects. The form of many plants had evolved to the point whereby they were adapted to be fertilized by one, or a very few, types of insect. Thus, while cowslips could be fertilized by either bumble-bees or moths, primroses depended almost entirely upon moths. In di- or trimorphic species the forms of the different flowers were such that different parts of the same insect might be used to brush off, and carry away, the pollen from different morphs. The fates of plants and insects had become inextricably bound together by their evolutionary histories.

Thus far, Charles's studies of the form and function of flowers were an unqualified success, bringing clarity where there had been only confusion. He had demonstrated not only the value of out breeding but also the floral mechanisms used by plants to maximize its occurrence. In addition, he had distinguished the form of insect pollinated flowers from that of wind pollinated ones. Charles invested several more years of his life in an attempt to define what fertility barriers exist, or do not exist, between species and varieties. Transferring pollen between flowers and waiting for seed to set he found the results of inter-specific crosses were unequivocal; offspring were sickly, or infertile, or both, just as were those which, on rare occasions, he obtained from 'illegitimate' intra-specific crosses. Unfortunately, in spite of his prodigious efforts, involving many plant species and thousands of test crosses, his progress was limited. He was able to formulate only a few simple generalizations because he was ignorant of the laws of heredity; he lacked the ground plan that would have enabled him to escape from the intellectual cul-de-sac into which he had wandered.

Convinced that 'Nature abhors perpetual self-fertilization'[42] and that most organisms have developed elaborate anatomical, morphological, or behavioural mechanisms to maximize out-breeding and minimize in-breeding, although he recognized that breeding directed by man could sometimes circumvent such barriers, and clear in his mind that out-breeding generated the variation upon which Natural Selection operated, Charles failed to recognize that discrete particles were the currency of inheritance. For perfectly understandable reasons, given his fixation on fecundity and fitness, he looked always at the vitality and fertility of the progeny of his test-crosses, or at the frequency with which bees visited their flowers. He was blind to the inheritance of simple, discrete characters, such as the colour or smoothness of seeds, as used by Gregor Mendel in his classic study of inheritance in peas; he failed to see that such characters were re-assorted according to simple mathematical rules when out-breeding (sexual reproduction) occurred. He was led, finally, to espouse the doctrine of pangenesis. According to this theory, every separate part of the whole organism reproduces itself. Thus, ovules, spermatozoa, and pollen grains – the fertilized egg or seed, as well as buds – include and consist of a multitude of germs thrown off from each separate part or unit. Pangenesis, like many other ill-founded theories about inheritance, was finally laid to rest in 1900 with the re-discovery of Mendel's writings and establishment in 1902 of modern genetics by, among others, William Bateson, Francis's friend (see Chapter 8).

*****

The Variation of Animal and Plants Under Domestication, containing those fragments of information that Charles had long stored in readiness for the 'big book', appeared in 1868 and only after Charles had fought a long battle against the tedium of the task. By that point his career had taken a sharp change in direction. While dealing with the extra workload associated with the avalanche of correspondence that *On the Origin of Species* generated and, also, the revisions and translations that publishers demanded from him, Charles had tired of the stresses and strains of the evolutionary debate and the constant need to rebut his critics; he needed a new, refreshing interest. In this receptive state of mind his enthusiasm had been captured by two of the most exotic groups of plants. In truth, he had long been interested in both insectivorous plants and orchids, collecting the former and studying at leisure their peculiarities. Now, however, these unusual plants vied for his attention. He was initially captivated by the insectivorous plants, but it was the orchids that eventually won. They were the first to receive his undivided attention.

The Orchidaceae is probably the largest family of flowering plants with nearly 30,000 species. A 'difficult' group to study, not so much because most species are tropical in origin and, therefore, require special conditions of light and temperature if they are to be grown in Britain, but more because their nutrition is specialized, and their flowering structures are complex and very different from those of most other plants. It may be wondered, therefore, why did Charles choose them? One reason may have been the very fact that the group was difficult and packed with secrets. Years after making his choice, he remembered that he had wanted to 'work out' a group of plants as carefully as he could, to learn everything about them. He had used the same highly focussed approach before with great success, when he had spent eight years in the 1840s and '50s studying barnacles (Cirripedia), and he wanted to employ such a highly focussed approach again. He always enjoyed a challenge and was eager to learn more about mechanisms of fertilization and fertility barriers in different groups of organisms. The orchids were ideally suited for this purpose. Thirty years earlier, conversations with Robert Brown about his study of the complexities of fertilization mechanisms in orchids had intrigued Charles and maybe he was still cogitating upon questions raised by Brown and their possible answers. Undoubtedly, one factor that helped attract him to the orchids was their beauty and local abundance. Down House was in those days deep in the Kentish countryside not swallowed as it is today by London's fringes. It was fine orchid country and a favourite afternoon walk for Charles and Emma was to a local beauty spot to which they gave a private name, Orchis Bank. Set in the quiet Cudham valley, the bank was a haven for both the Fly Orchid (*Ophrys insectifera*) and the green-flowered Musk Orchid (*Herminium monorchis*), as well as several Helleborines.

Charles went foraying for orchids among the local fields and lanes, digging up whole plants for the gardeners at Down House to maintain. He quickly learned the practical lesson that orchid plants needed to be transferred together with a good volume of their native soil if they were to flourish in the garden. The explanation, discovered years after his death, is that orchids have mutualistically symbiotic relationships with soil fungi. Orchids produce tiny seeds, each of which stores only the smallest quantity of nutrients, so the developing plant is heavily dependent on nutrients supplied by the fungus, a relationship maintained to maturity in many orchids. (Charles was well aware that orchids demonstrate the Malthusian principle that in biological species producing large numbers of poorly endowed offspring most perish before maturity).

With donations of plant material from friends such as Hooker, he was able to build an extensive collection. He embarked first upon a typically detailed study of fifteen genera of British orchid. What he found confirmed this was an exceptional family, charming and eccentric, whose basic floral parts were extravagantly adapted to attract bees, butterflies and moths before guiding them to nectar and pollen. He found pollen masses that were violently shot away from the flower, up to a distance of one metre, when he probed the flower with a pencil or a paintbrush. He asked, what is the significance for insect pollination of such violent ejection? And, what floral structures are necessary to ensure cross fertilization? To find answers, he began with familiar temperate species. Only when he was confident that he understood their relatively simple structures did he move on to study the even more complex tropical species.

Whereas the first edition of *On the Origin of Species* contains only the briefest reference to orchids, by the sixth edition (1876) Charles was able to adduce detailed evidence from his own studies further to substantiate key arguments made in earlier editions, in particular, arguments that had not wholly convinced some of his readers. He was, for example, often asked, how is the *origin* of structures explained? Although their utility might be obvious when they were fully developed, it was often hard for readers to see the structures' usefulness in their 'incipient, infinitesimal beginnings'. Orchids provided excellent ammunition for his arguments.

The floral parts of orchids are in units of three (Figure 5.1), although that basic number is not easily observed in the flowers of most species, where complex structures peculiar to the orchids are seen. One such structure is the rostellum, whose function is to produce a viscid secretion that helps attach pollen to passing insects. Robert Brown had ventured the opinion that it was not a new structure but was instead a modified stigma. Charles supported his old friend's view, pointing out that the rostellum was absent in orchids, such as Cypripedium, with three stigmatic surfaces. It is a highly variable structure but wherever it occurs its venation is similar to the venation of each of the two remaining

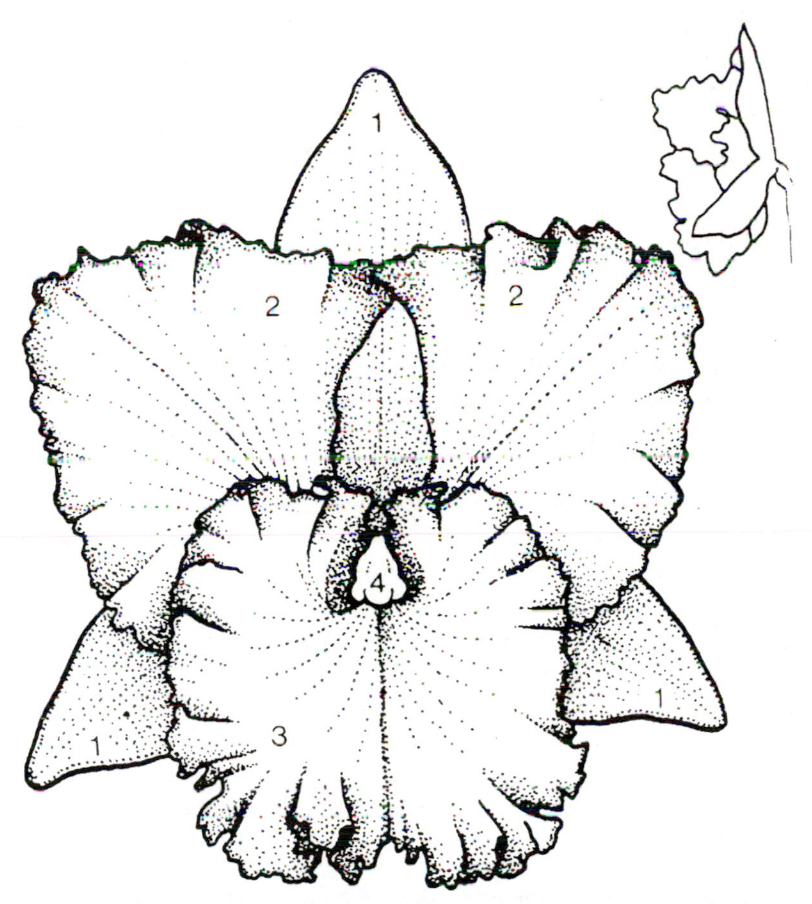

Figure 5.1: A simple orchid flower. In the genus *Cattleya*, three sepals (1) and three petals (2) are easily recognized. The lower petal (3) forms a lip or labellum, which is a landing platform for pollinating insects, while the finger-like column (4) contains both the male and female reproductive parts – the pollinia and stigmatic surface, respectively. Reproduced by permission of Anova Books, London from B. Williams, *Orchids for Everyone* (London: Treasure Press, 1984).

stigmas. Charles saw here clear evidence for the evolution of a complex structure out of a simpler one.

Another complex structure for which, he argued, there was evidence of its evolution from more simple structures was the pollinium. When highly developed, a pollinium is made up of a mass of pollen grains attached to an elastic foot-stalk, or caudicle, which, in turn, is affixed to a small globule of sticky, viscid material secreted by the third stigma. The caudicle sticks to (or in some instances is actively projected onto) an insect visiting the orchid flower and by this means pollen is transferred between flowers. But how did this highly developed complex of structures and materials evolve? In some orchids, there is no caudicle and the grains are merely tied together by fine threads, while in others the threads cohere at one end of the pollen mass, suggesting the beginnings of a caudicle. The aborted pollen grains sometimes found embedded within the central and solid parts of species with well developed caudicles were, Charles thought, further evidence of the origin of a complex structure from simple beginnings. He wrote,

> If we could see every orchid that has ever existed throughout the world, we should find all the gaps in the existing chain ... filled up by a series of easy transitions.[43]

In passing, it is noteworthy that similar evidence of gradual development towards complex structures and behaviour was presented by Charles in relation to climbing plants. In *The Movements and Habits of Climbing Plants* (1865), which is discussed more fully in the next chapter, he divided climbing plants into [stem] twiners, leaf climbers and those provided with tendrils. Twining depends on the curling of the young, flexible shoot around a support. Such twining, he pointed out, had arisen independently in widely separated species and families but it was not unique to climbers, for plants that do not climb show, in lesser degree, the same rotatory motion. Similarly, the sensitiveness of leaf stalks and tendrils to touch, which ultimately results in them curling around a support, is merely a well developed example of a more general phenomenon. He reported that his own tests with *Oxalis* and other non-climbers had clearly shown that they too moved in response to mechanical stimulation, albeit relatively insensitively.

To return to orchids, Charles attempted to demonstrate that natural selection may in different organisms achieve the same end by different ways. He focussed on nectar and the nectarines of orchids. Fundamental to Charles's theory of evolution was his belief that:

> it is an almost universal law of nature that the higher organic beings require an occasional cross with another individual; or, which is the same thing, that no hermaphrodite fertilizes itself for a perpetuity of generation. Having been blamed for not propounding this doctrine without giving ample facts, for which I had not sufficient space in that work [*On the Origin of Species*], I wish here [*Orchids*] to show that I have not spoken without having gone into details.[44]

His starting point was the non-contentious view that both in species having separate sexes, and in hermaphrodite species, pollen does not fall spontaneously on the stigma, some aid being necessary for fertilization. The simplest plan is for the light and tiny pollen grains to be blown from one flower or plant to the stigmatic surface(s) of another. Alternatively, and almost as simple, is the secretion of nectar to attract insects, which carry pollen between flowers and plants.

> From this simple stage we may pass through an inexhaustible number of contrivances, all for the same purpose and effected in essentially the same manner, but entailing changes in every part of the flower. The nectar may be stored in variously shaped

> receptacles, with the stamens and pistils modified in many ways, sometimes forming trap-like contrivances, and sometimes capable of neatly adapted movements through irritability or elasticity.[45]

Charles was especially fascinated by the mechanisms used by two tropical orchids to harness insects to achieve cross-fertilization. The first, *Coryanthes*, he learned about from the writings of Dr Hermann Crüger, Director of the Trinidad Botanic Garden; the second, *Catasetum*, he was able to study at first hand after he had begged Hooker for specimen plants. In the gigantic flowers of *Coryanthes*, part of its labellum, the lower petal or lip upon which visiting insects alight, is hollowed out into a great bucket into which drops of almost pure water continually fall from two secreting horns which stand above it (Figure 5.2). The mid- part of the labellum almost closes the bucket, leaving only two lateral exits; it also forms a chamber in which there is a series of fleshy ridges that are extremely attractive to humble bees. Large numbers of bees crowd into the chamber in order to gnaw the ridges, often knocking each other into the bucket below. Wet and temporarily flightless, they are thus compelled to crawl to one of the exits, guided by a set of strategically placed hairs. In doing so they rub first against the viscid stigma and second against the viscid pollen masses. With the pollen firmly attached to its back, each bee visits further flowers whereupon it inadvertently deposits the pollen on the stigmatic surface. By passing the stigmatic surface before reaching the pollen, the bee is thus unlikely to effect self-pollination of the flower. The flower is in exact harmony with its pollinator.

Bees visit the flowers of *Catasetum* to gnaw the fleshy labellum, just as they do in *Coryanthes*, but the overall structure of the flower of the two species is very different. Indeed, in *Catasetum* there is more than one flower type for the sexes are separate. The male flower contains a long tapering and sensitive structure, called by Charles the 'antenna'. When moved, as by a bee, the antenna triggers a spring that shoots the pollen mass outwards onto the back of the insect. The mass is transferred by the insect vector to the next female flower that it visits. *Coryanthes* and *Catasetum* are quite closely related and both use bees to effect pollination, but the two orchids have developed very different structures in order to achieve the same end, cross-fertilization.

Apart from underlining the normality of sexual reproduction, Charles's discovery that *Catasetum* had separate male and female flowers provided him with a minor triumph over the so-called orchid experts who had held that otherwise similar plants bearing obviously different flowers should be placed in two distinct genera – the male being called *Catasetum*, the female *Monocanthus* – with the occasional hermaphrodite flower being placed in a third genus. John Lindley, an orchid expert to whom Charles had sent his book, proclaimed in panic, 'such cases shake to the foundation all our ideas of the stability of genera and species'.

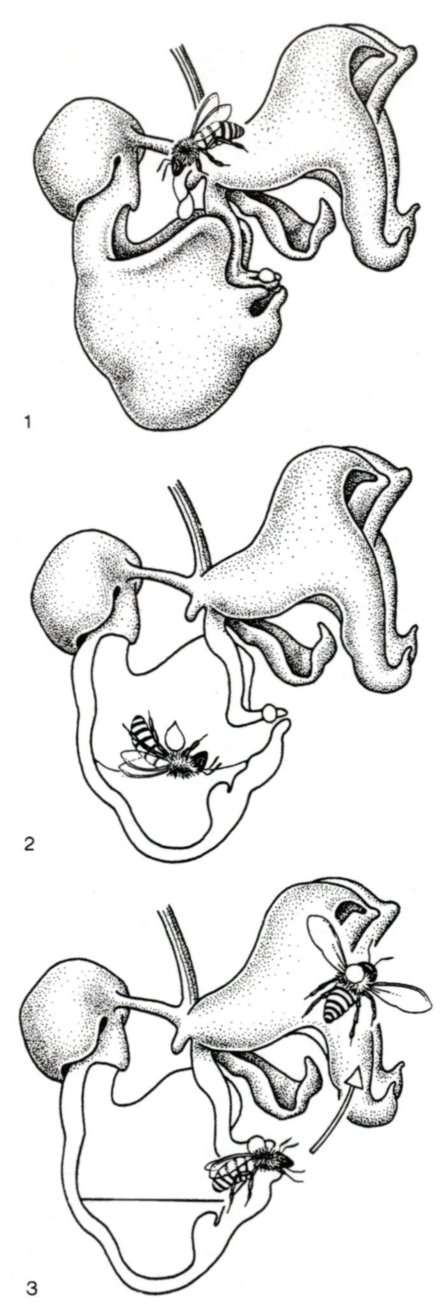

Figure 5.2: Pollination in *Coryanthes macrantha.* (1) Bees are attracted to the flower by a fleshy lip, on which they gnaw, but (2) they often slip or are knocked into the liquid-filled bucket below. (3) In exiting, the bee has to push hard on the lip above, collecting pollinia on its back. Reproduced by permission of Anova Books, London from B. Williams, *Orchids for Everyone* (London: Treasure Press, 1984).

Charles's painstaking dissections and observations had once again brought order to chaos. He had found in orchids a system whose principles closely resembled those he had found in barnacles so long ago. In short, organisms in each group were hermaphroditic, although functionally individuals were either male or female thanks to a complex series of structural changes having occurred to modify the basic hermaphrodite towards the production of either male or female organs. Charles's basic tenet was supported: out-breeding is the norm: Nature abhors perpetual self-fertilization. Some questions remained. Why did in-breeding persist at low levels? Why did some species not produce nectar? In the main, however, Charles had taken understanding of orchid biology to a new level.

Orchids was completed quickly and published in 1862. Largely descriptive, it follows in its organization Lindley's division of British orchids into five tribes. It is the last two chapters, 'Homologies of the flowers' and 'Miscellaneous and general considerations', respectively, that draw out the more general lessons and provide material for the later editions of the *Origin*. To Charles's delight, Hooker, Gray and a host of botanists heaped praise on his book. It instantly became the standard reference work on orchids. Charles's only regret was that his first purely botanical book, of which he was understandably proud, could not be read by his old mentor, Henslow, who had died a year before its publication.

*****

The marvellous floral adaptations that Charles studied for over forty years bore one final fruit. *The Effects of Cross and Self-Fertilization in the Vegetable Kingdom* (1876) was Charles's last book on the subject of fertility, selective advantage and heredity. Intended to be a complement to *Orchids*, the book was exceedingly wide ranging, exploring the possible origins of most of the world's major food crops, each family, or small group of families, having its own dedicated chapter. It stands as an excellent record of the state of knowledge at the time but offers little else to the modern reader.

In the weeks and years immediately following publication of *On the Origin of Species*, orchids and insectivorous plants had competed for Charles's attention. Orchids had won the day but with that project complete it was now time for him to turn to insectivorous plants.

6 CHARLES DARWIN'S PHYSIOLOGICAL PERIOD

To locate the end of Charles's evolutionary period and the beginning of his physiological period as occurring precisely in 1860, as Francis did,[1] suggests that a sudden change overtook his father. No such abrupt change happened, of course, although Francis's essential message was nevertheless valid; a watershed had occurred in 1860 marking a fundamental change in his father's focus and method of working. There was, in reality, a slow and lengthy period of transition, full of hiccups. Throughout the 1860s and '70s, Charles switched the focus of his attention from one group of plants to another, and then sometimes back again (see Table 6.1, overleaf). Projects overlapped with one another. First he would be concerned with insectivorous plants, next with orchids, then again with insectivorous plants – with attention to climbing plants being fitted in between. Capable of a seemingly obsessive interest in one group, he had an astonishing ability to keep several other interests ticking-over in the background. Whether it was by observation of the plants in his garden or greenhouse, or those in the countryside around Downe, he was always watching and noting what was happening to plants, often tinkering with them in ways that would one day lead to major investigations. Prize plants, sent by Hooker or some other botanical correspondent, were often kept on the windowsill in his study where he could watch them day by day, or even hour by hour, making notes that might be useful at some time in the future.

From his studies of climbing plants and of insectivorous plants he learned new facts about convergent evolution and about competition, facts that complemented his evolutionary studies, providing additional material for his regular revisions of *On the Origin of Species*. Little by little, however, his viewpoint changed. He became more and more aware of questions about possible links between structure and function. His observations of the circular movements of the shoot apices of climbing plants, for example, or the rapid movements of the sticky tentacles and snapping leaflets of insectivorous plants, led him to contemplate the physiology of the plant. Gradually, rather than suddenly, he was asking himself questions about plant functions. It was forcefully brought home to him what his great friend Huxley had always preached. Namely, function is an end result that emerges when all the components of a system are assembled in a particular way, albeit by chance. A study of function – Huxley was among the first to call it physiology – was enlightening for it enabled the facts of morphology to be deduced.[2]

Table 6.1: Charles Darwin's Major Interests in Plants (Modified from T. Veak, 'Exploring Darwin's Correspondence; Some Important but Lesser Known Correspondents and Projects', *Archives of Natural History*, 30 (2003), pp. 118–38).

Topic	Most intense period of study	Publications, with notes
Seed dispersal and viability	1855–67	Information incorporated in later editions of *On the Origin of Species*.
Flowers and fertilization	1860–77	(1) *The Different Forms of Flowers on Plants of the Same Species* (1877): a collection of five papers previously read to the Linnean Society between 1861 and 1868. Republished in 1884 with a Preface by Francis Darwin evaluating progress in the subject since 1877. (2) *The Effects of Cross and Self Fertilization in the Vegetable Kingdom* (1876; 2nd edn, 1878).
Orchids	1860–77	*The Various Contrivances by which Orchids Are Fertilised by Insects and the Good Effects of Intercrossing*, (1862; 2nd edn, 1877, with '*and the Good Effects of Intercrossing*' omitted from the title).
Climbing plants	1863–5	*The Movements and Habits of Climbing Plants* (1865; 2nd edn, 1875).
Insectivorous plants	Briefly, 1860: mainly, 1872–5	*Insectivorous Plants* (1875; 2nd edn, 1888, edited by Francis Darwin incorporating his father's marginal notes and adding extensive footnotes of his own).
Irritability and movement in plants	1873–80	*The Power of Movements in Plants* (1880).

Ever since *On the Origin of Species* was published there has been an ongoing argument as to whether or not Charles was a teleologist.[3] Charles's approach to function is at the heart of the debate, for teleological arguments employ the concept of purpose or final cause. They are highly dangerous for implicit in them is the risk that they can be misused, as by those wanting to justify a belief in 'Intelligent Design'. Despite this risk, such arguments are at their best often extremely useful, not just because they can stimulate creative investigation but also because they can integrate in a sensible manner a number of apparently disparate facts or statements, so drawing out function, or cause and effect. To give an example, the evolution of each of the numerous anatomical features that regulate the movement of water through individual parts of the plant has its own intrinsic interest but can be fully understood only in the context of a combined purpose – that of controlling water flow through the whole plant. Teleology is thus opposed to reductionism. Indeed, one of the current bandwagons of biology, systems analysis, has at its core the idea that function is an emergent (or otherwise hidden) property of a system, requiring a description of parameters beyond those of the system's components.[4]

The Movements and Habits of Climbing Plants, even more so than *On the Various Contrivances by which British and Foreign Orchids are Fertilized by Insects* (see Chapter 5), abounds with teleological statements such as

> The first purpose of the spontaneous revolving movement, or, more strictly speaking, of the continuous bowing movement directed successively to all points of the compass, is ... to favour the shoot finding a support.[5]

Modern opinion favours the view that Charles was a teleologist, for whom such arguments were a useful tool, rather than a misleading distraction, as he became more and more involved in physiology.[6] By dealing with living things as though they had 'goal-orientated' actions, he could move more easily into mechanistic speculations about the physical and biological laws at work.

This chapter covers the first phase of Charles's physiological period, from 1860 to about 1875–6, when he was still working mainly on his own; they were the years before Francis joined him as amanuensis and then collaborator. Underlying Charles's interest in climbing and insectivorous plants in this first phase were three physiological topics; the irritability (sometimes called the excitability) of plant tissues; the circumnutation (circling) and growth of stem apices; and nutrition. At the time, Charles recognized some of the connections between the three, while contemporaries such as Dutrochet and Pfeffer saw others. But it was only a little later – indeed, in the year of Charles's death, 1882 – that John Burdon Sanderson was able to coordinate the different men's ideas, setting out a unified picture that is largely recognizable today. Each of the threads that led to Sanderson's synthesis will now be traced.

*****

Virginia Creeper and clematis still clamber over the walls and veranda of Down House, just as they did when Charles was alive, and in Kent there are still vestiges of the hop bines that in his lifetime filled its fields. Charles may have been surrounded by climbing plants every day but he only began tinkering with them after he had read in 1858 a short paper published by Asa Gray on the subject of the movements of tendrils of some cucurbitaceous plants.[7] At first his investigations were constrained by a lack of growing space but in the spring of 1863 completion of a new heated greenhouse, built against the wall of the kitchen garden at Down House, enabled him to grow the large numbers of climbing plants that he would need as he devoted more and more time to describing and measuring the circling movements of their stems, leaflets and tendrils.[8] His enthusiasm was rewarded as he made rapid progress, but then Gray dropped a potential bombshell. Aware of what Charles was attempting, he warned him about two little known but prize-winning essays on the subject of climbing plants submitted to the medical faculty of the University of Tübingen by Ludwig Palm and Hugo von Mohl, and both published in 1827.[9] The unavoidable truth was that the two German investigators had long before described the phenomenon of rotating stems and tendrils. What should Charles do? Had he wasted his time? After much deliberation, he felt his own observations had 'sufficient novelty' to justify publication. He would, he decided, go on. So, he completed the project, despite meanwhile learning of two more memoirs published on the subject of climbing plants, this time by Henri Dutrochet in the French journal *Comptes Rendus* (1843; 1844), and both, in Charles's opinion, excellent.[10] After making hundreds of meticulously recorded observations, and much self-questioning, Charles finally wrote up his findings in an 138-page long paper published in the *Proceedings of the Linnean Society* (1865). At the same time the society's publishers, Longmans, produced the article as a 'separate', a chunky pamphlet which was effectively the first edition of the book. While acknowledging the works of earlier authors, Charles, as usual, concentrated on his own results. Ten years later, in 1875, and after further shorter periods of study, a second edition of *The Movements and Habits of Climbing Plants* was published by John Murray.

Plants become climbers, he proposed, in order to keep their leaves above those of their neighbours. By this means they avoid shading, maintaining their leaves in full sunlight and exposing them to free, moving air. In comparison with trees, climbing plants can thus attain similar heights, but with the expenditure of very much less 'organized' or dry matter, so they are at a competitive advantage in the race to sunlight.

As he gained experience, Charles saw in climbers a taxonomically diverse group of plants, but with the same habit of climbing, that had evolved a few common mechanisms, including twining of stems, or of tendrils or leaves, to achieve their ends. Hook climbers, such as the dog-rose, were the least efficient, he concluded, since they can climb only in the midst of an entangled mass of vegetation, while root climbers, such as ivy, although excellently adapted to ascend the bare faces of rocks or the trunks of trees, were comparatively rare. Much more adaptable, and therefore widespread, were the two great classes of climbers which he recognized, the stem twiners, such as the hop, whose shoot apices continually describe circles as they extend, exploring for objects around which they can twine, and those plants which have special sensitive twining organs, either leaf climbers, such as *Clematis* spp., or tendril-climbers, such as Passiflora.[11] Most of this last type still had rotating *shoot* tips but, thought Charles, they had devolved the specialist task of attachment to leaves or tendrils, their stems being allowed to thicken – thick stems being relatively poor at wrapping themselves around other stems and branches – so that each plant can support a much greater number of the attaching organs, whether leaves or tendrils.

Having settled the general principles of the climbing habit, Charles moved on to more detailed investigations, first using the same plant, *Echinocystis lobata* (wild cucumber), that had fascinated his friend Gray. Following a familiar pattern, it was another friend, Hooker, who supplied Charles with specimens of this tendril-climber from Kew. Supplies were, however, limited so Charles soon switched his attention to the hop, which was much more readily obtained locally. *Climbing Plants* begins with a lengthy and detailed account of the rotations of this twining plant as it curled around a variety of objects, including Charles's walking stick. He progressed to comparing a range of twining species, tabulating their rates of revolution at different dates and times of day, and the thickness of cylinders around which they would, or would not, circle. He made scores of permanent traces of the tentative circles of shoot apices.[12]

Charles and Francis later gave the rotating habit a special name, 'circumnutation', much to the annoyance of Julius von Sachs who, in his *Lehrbuch der Botanik* (1865), had simply called it 'nutation'. (Sachs would be pleased to know that posterity has adopted his shorter name.) A further century or more of research has revealed much about nutation. Cell division in plants is restricted to discrete regions or zones called meristems, those in shoots and roots being located just beneath the surface of their tips. As parent cells divide to form new cells, the more posterior ones increase in volume with the result that the tip of the shoot or root extends progressively forward. It seems that nutation involves a wave of cell extension growth moving around the tip; at any instant, one region of the tip is relatively quiescent while there is activity in the remainder. The wave of activity passes around the complete tip in a cycle lasting from 1.25 to 24 hours according to species. Most but not all shoot apices rotate in an anti-clockwise direction when viewed from the top.[13] It has been suggested that nutation is a special case of gravitropism (the response to gravity in which stems bend upwards, while roots bend downwards – see the next chapter). The idea is that as a stem reaches a vertical orientation, in response to gravity, it may continue to grow more rapidly on what was until recently the lower side, so passing the vertical before the opposite gravitropic response takes over – a form of overshoot that causes the apex to describe circles or ellipses. The remarkable truth is that even today the basic mechanism underlying nutation is still not fully understood.

Despite *Climbing Plants* being the most comprehensive and accurate account of the group written at the time, the book sold poorly. Perhaps climbing plants were too familiar to practical botanists, lacking the mysterious and exotic properties of orchids, while, for theorists, the book merely consolidated well trodden paths rather than opening up new ones. Although still largely accurate and correct, and most importantly leading Charles to a study of tropisms (see Chapter 7), the book's interest for the modern reader lies as much in its continuing defence of *On the Origin of Species* as in its subject matter per se. Climbing plants, he argued, provided excellent evidence of the way in which different structures can evolve in distantly related organisms to achieve the same end. In the concluding pages of *Climbing Plants*, Charles poses questions:

> The divisions containing twining plants, leaf-climbers, and tendril-bearers graduate to a certain extent into one another, and nearly all have the same remarkable power of spontaneously revolving. Does this gradation ... indicate that plants belonging to one subdivision have actually passed during the lapse of ages, or can pass, from one state to another? Has, for instance, any tendril-bearing plant assumed its present structure without having previously existed as a leaf-climber or a twiner?[14]

Answering his own questions, he replies:

> the idea that [leaf-climbers] were primordially twiners is forcibly suggested. The internodes of all, without exception revolve in exactly the same manner as twiners; some few can still twine well, and many others in an imperfect manner.
>
> ... it is probable that all tendril-bearers were primordially twiners, that is, are the descendants of plants having this power and habit. For the internodes of the majority revolve: and, in a few species, the flexible stem still retains the capacity of spirally twining round an upright stick. Tendril-bearers have undergone much more modification than leaf-climbers; hence it is not surprising that their supposed primordial habits of revolving and twining have been more frequently lost or modified than in the case of leaf-climbers.[15]

Always conscious of the need to rebuff potential criticisms, he added:

> We have seen that tendrils consist of various organs in a modified state, namely, leaves, flower-peduncles, branches, and perhaps stipules ... But he who believes in the slow modification of species will not be content simply to ascertain the homological nature of different kinds of tendrils; he will wish to learn...by what actual steps leaves, flower-peduncles, etc., have had their functions wholly changed, and have come to serve merely as prehensile organs.[16]

And replied to himself:

> evidence has been given that an organ, still subserving the functions of a leaf, may become sensitive to a touch, and thus grasp an adjoining object. With several leaf-climbers the true leaves spontaneously revolve; and their petioles, after clasping a support grow thicker and stronger. We thus see that leaves may acquire all the leading and characteristic qualities of tendrils, namely, sensitiveness, spontaneous movement, and subsequent increased strength. If their blades and laminae were to abort, they

> would form true tendrils. And of this process of abortion we can follow every step, until no trace of the original nature of the tendril is left.[17]

This process is then traced from *Mutisia clematis,* where the tendril closely resembles the leaf midrib but there is still a vestigial lamina, through four genera of the Fumariaceae, where the terminal leaflet is progressively reduced, to the Dicentra where the petiole is perfectly characterized.

> In *Tropaeolum tricolorum* we have another kind of passage; for the leaves which are first formed on the young stems are entirely destitute of laminae, and must be called tendrils, whilst the later formed leaves have well-developed laminae. In all cases the acquirement of sensitiveness by the midribs of the leaves appears to stand in some close relation with the abortion of their laminae or blades.[18]

*****

Charles's next consuming botanical interest was insectivorous plants, but before considering what he did with them, some context is helpful. To this end, it is necessary to take a step backward, picking up a different thread of research that ultimately led to the synthesis made by Burdon Sanderson. This new thread involves the same man, Dutrochet, who had studied the periods of rotation of tendrils in the garden pea under the influences of temperature and of light, which Charles had so admired. In this instance, it is Dutrochet's discovery (some would say rediscovery)[19] in the late 1820s of the phenomenon of osmosis that is of interest. The thread involves, also, Wilhelm Pfeffer, who realized and demonstrated the relevance of osmotic phenomena to movements in plants.

In 1809, the sickly Henri Joachim Dutrochet resigned his commission as a surgeon in the French army and at the early age of thirty-three retired to his family's estate at Chareau, in Touraine, which his mother had managed to retain through the Revolution (unlike much of the property owned by his aristocrat father, the Comte du Trochet). To occupy his restless mind, he turned to the natural sciences and, like Charles Darwin in the peaceful seclusion of his country house at Downe, Dutrochet became an outstanding biologist. In 1824 he published his first book on the microstructure of plants and animal, in which he supported the contemporary German view that tissues were comprised of many globules or vesicles, rather than the French view that tissue was a continuous fabric full of holes, rather like a string vest!

Like Charles, Henri Dutrochet too was extremely reluctant to let travel disrupt the quiet routines of country life, although he did make an exception when in 1827 he left France to visit Andrew Knight in England.[20] He admired Knight greatly and perhaps through their regular correspondence he had been assured that at his host's country estate in Herefordshire he would find the peace and tranquillity familiar to him at Chareau. Knight was primarily a horticulturalist whose views on out-breeding were well known to Charles (see Chapter 5), but he also dabbled in plant physiology – usually with great distinction as when (see Chapter 7), he described the gravitational responses of root tips. Dutrochet was enthused by his host's interest in physiology, including that in water relations, and, on returning to France, was inspired to repeat several of Knight's experiments.

Some years before that visit to England, Dutrochet's earliest biological studies had concerned the development of eggs in birds, reptiles and mammals. Plant movements (including closure of leaves of the touch-sensitive plant, *Mimosa*) had caught his attention, as had the work of Stephen Hales, many of whose experiments on the movement of sap in plants he had tried to repeat. In 1826 he was examining sperm sacs in snails when he saw one of them burst open in water. He could see no contraction of the sac; the sperm seemed to be simply forced out as water came in from behind. He had seen a similar phenomenon when looking at bursting sporangia of the fish-infecting fungus, *Saprolegnia*, and was intrigued by the possibility that there was a common explanation for both these events, and for such diverse phenomena as Hales's root pressure and the ascent of sap in plants. He made an hypothesis:

> The presence of a substance denser than water inside small organic cavities, is thus one of the necessary conditions bringing about the physio-organic action which violently introduces water into their interiors.[21]

To test the hypothesis he took a short length of animal intestine and filled it with milk. Submerged in water, it swelled up. Here was a new force which he called 'endosmose'. If plants (and animals) were made up of tiny vesicles, filled with fluid, endosmose would explain, he thought, the motions characteristic of their lives. And if some form of vital motion required a stimulus as well as fluid flow, then there must be additional nervous action, which was known to be an electrical effect. His experiments were crude and not very accurate but he was able to construct a simple osmometer with which he demonstrated that if a natural membrane separated aqueous solutions of different concentration of the same substance, there was a net flow of water across the membrane from the concentrated side towards the dilute side, while the dissolved substance (the solute) did not pass freely from one side to the other. Dutrochet had struck a blow in the war against the existence of a vital force in living organisms; he had added to the growing list of biological phenomena that could be explained simply by applying the laws of physics and chemistry.

Later, Dutrochet compared the osmotic effects of salts and sugars, and tested the semi-permeability of membranes ranging from onion scales to the bladders of domestic animals. He recognized the curling that occurs in freshly excised plant tissues when they are placed in water – and which is reversed by bathing them in concentrated salt solutions – has its origins in the osmotically driven redistribution of water between the cells of the tissue and their external solution.[22] In practice, movements are limited by physical adhesion between cells and, in addition, whether or not the solution within cells is able to reach the same concentration as the external solution (to achieve osmotic equilibrium). The internal spaces and cell walls of isolated plant tissues are penetrated by, and often reach osmotic equilibrium with, any bathing solution.

Among his other distinctions, Dutrochet asserted that respiration is of the same nature in plants and animals, showed that the stomata in leaf surfaces connect the outside air with lacunae, air chambers within the leaf, and, as seen already, described nutation in shoot apices. But it is for his discovery of osmosis that he will always be remembered. Seized upon with enthusiasm by German botanists, his discovery of osmosis was slow to impress either his fellow Frenchmen or the English. Thus, de Candolle did not mention osmosis in his books. To his credit, Henslow did touch on the subject of osmosis in *The Principles of Descriptive and Physiological Botany* and even used one of his rare diagrams to show an apparatus – two enclosed solutions separated by a membrane – in which osmosis could be demonstrated, but he went no further. Unfortunately, neither Henslow nor his student Charles Darwin recognized the significance of osmosis and cell water relations to plant movements. Most notable among the German biologists who were alert to the importance of osmosis was Wilhelm Pfeffer. Born near Kassel, in Germany, but the son of a Swiss pharmacist, the young Pfeffer had studied both chemistry and physics at Göttingen.[23] Equipped with this unusual blend of experiences and skills, it was he who in 1877 took the important step of quantifying the osmotic pressure (today celled osmotic potential) of plant cells.

His measurements were the culmination of a long trail of experiments and planning driven by an interest in those movements of the leaf of the sensitive plant, *Mimosa*, that follow touch or an equivalent physical stimulation, and which were well known to both Erasmus and Charles Darwin. Pfeffer observed that as the leaflets of detached leaves collapsed there was an accompanying secretion of a drop of sap from the cut leaf stalk. Pfeffer concluded that closure was due to a sudden rush of water out of 'motor cells' which, thereby, lost their turgescence. But the anatomy of *Mimosa* leaves is complex, for like all leaves they contain several different tissue and cell types, making easy interpretation and further progress difficult. If understanding was to advance then, Pfeffer realized, a simpler system was needed. He chose a much simpler one, the anther filaments of the stamens of *Centaurea jacea* (knapweed). A member of the Compositae, knapweed's flowers are small (= florets) and are aggregated into heads or capitulae. At the centre of each floret, the style is surrounded by a tube of five fused anthers, broken only in the lower region where the five filaments separate and arch outwards. When mature, the filaments rapidly straighten if any one of them is touched, causing the style to be drawn down to the anther tube. Accepted theory was that straightening was comparable to the contraction of a muscle but when Pfeffer made exact measurements of the changes in length and thickness of the filaments he recognized that their *volume* decreased. The reduction in volume coincided with the appearance of a drop of sap on the surface of a cut filament; the drop was reabsorbed, by what Pfeffer called 'endosmosis', as the filament later relaxed to become arched once more.

Pfeffer concluded that the movements of the anthers were due to osmotic forces.[24] Next he attempted to quantify those forces. He hung weights from contracting filaments in order to find out what 'force' was needed for expansion to take place without osmosis. The values were much higher than what seemed possible in the light of current knowledge, forcing Pfeffer to reconsider the basic physical phenomenon. His first step was 'to determine what osmotic pressures are exerted by dissolved substances, and especially by the so-called crystalloids, when they do not diosmose [do not pass through the membrane]'.[25] After many trials and much criticism, and using a porous pot on the inside of which was precipitated an artificial membrane of copper ferrocyanide, he was able to measure accurately osmotic pressures. From these new measurements it was clear that his earlier ones had indeed been correct. Even the relatively dilute solutions extracted from plant cells could generate osmotic forces sufficient to drive phenomena such as the movements of leaflets of *Mimosa* or of anther filaments of knapweed. Pfeffer had thereby established the important principle that movements of plant tissues and organs can occur when the volume of critically positioned cells changes.

It is normal for the solution inside a plant cell to be more concentrated than that on the outside. Thus, root cells have solutions more concentrated than the soil solution surrounding them (except in conditions of severe drought), with the result that water moves into the plant from the soil. Leaf cells, which lose water to the *relatively* dry atmosphere, have more concentrated solutions than root cells, causing water to move up the plant or, as first proposed by Hales, to be pulled up the plant (see p. 39). In practice, the movement of water into plant cells is restricted by the cell wall that surrounds them. As they become progressively more turgid, either the cell wall stretches, allowing an increase in the volume of the cell and further water uptake, or water is stopped from entering as 'full turgor' is reached.

Pfeffer had pointed the direction to be followed if the connections between osmotic potential, turgor potential and the expansion of plant cells during growth of fresh weight were to be elucidated. However, significant further progress had to await a much better understanding of the biophysics governing the stretching of the non-living, cellulose-based wall that encases each plant cell. Laid down in cell division, i.e. during the 'birth' of the cell, the wall is relatively elastic in young cells and is easily stretched whenever the cell has sufficient turgor. To some extent, as in rotating shoot apices where new cells are continuously being formed within the apical meristem, the plant controls turgor, and therefore expansion, by actively pumping solutes into or out of cells. Biochemical changes in the wall cause it to lose its elasticity, most commonly during cell ageing. In these circumstances the wall usually thickens and the turgor potential is no longer sufficient to drive expansion; this is normally the stage at which the cell has reached its final, mature volume and shape.

In a very few specialized tissues, such as those involved in the movements of leaves or leaflets, rapid changes in membrane permeability of a few specially positioned cells, which may have retained thin flexible walls, leads to a rapid flux of solutes (usually *out* of cells) leading to their loss of turgor. Groups of cells can act in concert if they are connected by plasmodesmata, cytoplasmic strands linking neighbouring cells through pits in their juxtapposed walls. In turgid tissues, each cell commonly presses upon its neighbours, and vice-versa. Thus, if one or a few cells rapidly lose turgor, these internal pressures are changed, leading to movement.

*****

One of the more extraordinary facts about flowering plants is that they include among their number about 600 carnivorous species. The habit has evolved independently within at least six subclasses, their common imperative being the need to supplement a nutrient-poor environment. Carnivorous plants are most common in wet, open habitats with impoverished soils, places where plants have no problem in acquiring carbon through photosynthesis but where growth would be limited by a lack of nitrogen, unless it were obtained from small animals. In many of these plants responsibility for nutrient uptake has passed from roots to leaves; leaves which have been modified to trap animals and then digest the nitrogen-rich proteins in their bodies. The animals most commonly trapped are insects – explaining why the plants are often called 'insectivorous' – but the diet may in fact contain a wider range of invertebrates. Thus, Charles, who was the first to make an extensive study of this group, called them insectivorous even though he described some that fed on crustaceans. It was Charles who proposed the carnivorous habit was an adaptation to nutrient-poor habitats when he recognized how their growth was promoted when nitrogen-rich materials were fed to them 'in captivity'. Although such feeding experiments supply much more nitrogen than the plant is likely to obtain in the field, and tend to overestimate the amount of animal nitrogen assimilated by the plant, the actual amounts taken in are considerable. Modern stable isotope techniques, used to analyse plants growing naturally without artificial feeding, show that in *Drosera* spp. up to 50 per cent of the plant's nitrogen can be derived from trapped animals.[26] The percentage varies according to the plant's circumstances but, in broad terms, it is inversely related to the nitrogen supply in their immediate soil environment.

Charles's interest in insectivorous plants started in the summer of 1860, was than put aside in the autumn while he started a concentrated study of orchids, and was resumed in 1872, finally ending in June 1875 with publication of his book, *Insectivorous Plants* (see Table 6.1). The short flurry of intense activity in 1860 began during a summer holiday near Hartfield in Sussex when his attention was drawn to the common or round-leaved sundew, *Drosera rotundifolia*, that abounds in the local heathland and, in particular, to the many insects entrapped by its leaves.[27] The sundew has a basal rosette of leaves, each leaf bearing up to a hundred glistening tentacles on its upper surface. Each tentacle secretes a sticky mucilage which traps small insects, subsequent digestion being completed when the tentacles coil around the prey. On one large leaf Charles found thirteen different insect species. Aware that the leaves of many plants cause the death of insects by fortuitous, passive trapping – as on the sticky scale leaves of buds of horse chestnut, *Aesculus hippocastrum* – his genius told him that something different was happening in the case of sundews. He collected a dozen specimens and carried them home, whereupon he fed them with living insects and was immediately impressed by the way the sticky tentacles of the sundew's leaves curled around and imprisoned the insects. *Drosera* was obviously adapted to trap insects. A quick test, in which he placed droplets of various nitrogenous and non-nitrogenous substances on the leaves, strongly suggested that nitrogen was the stimulant to which the tentacles responded and a connection was made with the recent discoveries of Lawes and Gilbert – and indirectly of Liebig and Boussingault too (see Chapter 4). They had shown that a shortage of nitrogen often limits plant growth, so, Charles inferred, sundews must trap insects in order to acquire a supplementary source of nitrogen. Letters on the subject of insectivorous plants were exchanged with his usual circle of correspondents and Emma jokingly told Lady Lyell, 'he is treating Drosera ... just like a living creature, and I suppose he hopes to end in proving it to be an animal'.[28] It would be another twelve years before Charles could again give his whole attention to his beloved *Drosera*, but perhaps she was not far off the mark for in 1875, after publication of the book, Charles told Asa Gray, '[Drosera] is a wonderful plant, or rather a most sagacious animal'.[29]

The world knew little about insectivorous plants before Charles published his book and he, as ever in his various botanical endeavours, liked to play down any previous knowledge that he had about plants. In only the second sentence of the first chapter of his book, he declares, '[in 1860] I heard that insects were thus caught, but knew nothing further on the subject'.[30] But was he once more being, ever so slightly, disingenuous? After all, long ago his grandfather, Erasmus, vividly described insectivorous plants in both *The Loves of Plants* and *Phytologia*.

Erasmus had been fascinated by their relationship with insects, even if, typically, he interpreted it from a medical standpoint.

> Queen of the marsh imperial Drosera treads
> Rush-fringed banks, and moss-embroider'd beds
> Redundant folds of glossy silk surround
> Her slender waist, and trail upon the ground;
> Five sister-nymphs collect with graceful ease,
> Or spread the floating purple to the breeze:
> And *five* fair youths with duteous love comply
> With each soft mandate of her moving eye.
> As with sweet grace her snowy neck she bows,
> A zone of diamonds trembles round her brows;
> Bright shines the silver halo, as she turns;
> And, as she steps, the living lustre burns.[31]

Erasmus's footnote reads,

> The leaves of this marsh-plant are purple, and have a fringe very unlike other vegetable productions. And, which is curious, at the point of every thread of this erect fringe stands a pellucid drop of mucilage, resembling an earl's coronet. The mucus is a secretion from certain glands, and like the viscous material round the flower stalks of silene (catchfly) prevents small insects from infesting the leaves. As the ear-wax in animals seems in part designed to prevent fleas and other insects from getting into their ears.[32]

James Edward Smith, whose *Introduction to Physiological and Systematical Botany* (1807) was on Henslow's recommended reading list, devoted four pages to *Sarracenia* and *Nepenthes*, two bog plants whose leaves form pitchers to collect water and, thereby, trap insects. Smith had asked, what is the purpose of such structures? He answered:

> An observation communicated to me in 1805, in the botanic garden Liverpool, seems to unravel the mystery. An insect of the *Sphex* or *Ichneumon* kind [???] ... was seen by one of the gardeners to drag several large flies to the *Sarracenia adunca*, and, with some difficulty forcing them under the lid or cover of its leaf, to deposit them in the tubular part, which was half filled with water ... Probably the air evolved by these dead flies may be beneficial to vegetation, and, as far as the plant is concerned, its curious construction may be designed to entrap them.[33]

Although ignorant of the mechanism, Smith raised the possibility that the plant benefits *nutritionally* by trapping insects. He suggested that *Drosera* and *Dionaea muscipula* (Venus Flytrap) similarly benefit. It is only in the second edition of *Insectivorous Plants*, published in 1888, that one of Francis Darwin's many footnotes cites precedence for a nutritive theory, pointing out that, 'In the *Quarterly Journal of Science and Art*, 1829, G. T. Burnett expressed his belief that

Drosera profits by the absorption of nutritive matter from the captured insects'. Francis does not mention Smith.[34]

Apart from any memory of Erasmus, and the huge intrinsic interest of insectivorous plants, were there other reasons why these plants attracted Charles's attention? One possibility, which has more resonance with his evolutionary period than his physiological period, is that they may have been seen by him as encapsulating the struggle for survival. Carnivory, which had evolved independently in taxonomically distant groups of plants, demonstrated how parallel evolution may occur in response to a common driving force. (Moreover, there must be a fine balance between predator and prey; in theory, the plant's trapping mechanisms might evolve to a point at which they become too efficient and threaten the very abundance of the insect prey.)[35] But, oddly, Charles makes little of this evolutionary aspect of the biology of insectivorous plants. What is more, he completely ignores a further complexity, the conflict of interests for the plant arising from the fact that most insectivorous plants are pollinated by insects!

By far the greatest part of *Insectivorous Plants* is devoted to *Drosera* and its adhesive traps, but later chapters deal with other genera, such as *Pinguicula*, the Butterwort, which has a similar adhesive trap, *Utricularia*, the bladderwort, which grows submerged in water and has a suction trap,[36] and, most importantly, *Dionaea* with its snap trap. The book is illustrated by George and Francis and, again and again, cites observations and experiments made 'by my son, Francis' who, it should be noted, was busy elsewhere during this period completing his medical studies and researching nervous functions in snails. But more of that later.

Concerning the trapping and digestion of insects, Charles conducted hundreds of tests in which he subjected sundew leaves to particles of different size and to a vast range of organic materials, most of which it would never encounter in nature. He supplied too chemicals in pure form, for each test meticulously recording the results in considerable detail. Sections of *Insectivorous Plants* read rather like a laboratory notebook – which was a new but significant change of style for Charles.

At the core of his work, however, there were simple and straightforward conclusions. In his own words:

> When an insect alights on the central disc [where tentacles are shorter], it is instantly entangled by the viscid secretion, and the surrounding [longer] tentacles after a time begin to bend, and ultimately clasp it on all sides. Insects are generally killed ... in about a quarter of an hour, owing to their trachaea [breathing pores] being closed by the secretion.[37]

> When an object, such as a bit of meat or an insect, is placed on the disc of a leaf, as soon as the surrounding tentacles become considerably inflected, their glands pour forth an increased amount of secretion. They do not secrete the ferment proper for

digestion when mechanically irritated, but only after absorbing matter, probably of a nitrogenous nature.[38]

When the tentacles become inflected ... the secretion not only increases in quantity, but changes its nature and becomes acid.[39]

The secretion ... completely dissolves albumen, muscle, fibrin, areolar tissue, cartilage, the fibrous basis of bone, gelatine, chondrin, casein in the state in which it exists in milk, and gluten which has been subjected to weak hydrochloric acid.[40]

Tentacles remain clasped for a much longer average time over objects which yield soluble nitrogenous matter than over those, whether organic or inorganic, which yield no such matter.[41]

The absorption of animal matter from captured insects/ explains how Drosera can flourish in extremely poor peaty soil.[42]

Today it is known that insectivorous plants typically secrete a cocktail of enzymes that breaks down the bodies of their prey. Key components are the proteases, enzymes that attack the linear protein molecules from either their chain-ends (exopeptidases) or middle (endopeptidases), liberating their constituent amino acids, and phosphatases, enzymes that act on lipids and amino acids to release the phosphorus which is only a little less important than nitrogen to the nutrition of insectivorous plants. The whole process of digestion was, however, a complete mystery when the first edition of *Insectivorous Plants* was published in 1875 and, although by the time of the second edition in 1888 knowledge was accumulating about the function of pepsin and acid in human digestion, knowledge of digestion in other animals and, particularly, by plants had progressed very little.[43] In 1877 Sydney Vines[44] had extracted for the first time an 'active principle' from an insectivorous plant, in this case an extract from the walls of *Nepenthes* pitchers that would attack insect bodies, but generally understanding of plant enzymes, digestive or otherwise, was minimal. Indeed, although the word 'enzyme' was occasionally used, it had not been defined. It was not until the end of the 1920s that the biochemical properties of enzymes began to be fully understood and the term 'enzyme' was used with consistency.

*****

Writing to Joseph Hooker back in 1862, Charles had told his old friend of the existence in *Drosera* of 'diffused nervous matter', which he thought was in some degree analogous in constitution and function to the nervous matter of animals.[45] What had prompted this extraordinary remark was his observation that when the gland at the tip of a tentacle was stimulated it passed a signal first down the tentacle to the basal part, which alone bent, and then:

> passing onwards, [the signal] spreads on all sides to the surrounding tentacles, first affecting those which stand nearest and then those farther off ... When a gland is strongly excited by the quantity or quality of the substance placed on it, the motor impulse travels farther than from one slightly excited.[46]

Prompted by the realization that in a plant he could observe both movement in reaction to a defined localized stimulus and the transmission of that stimulus, in the form of some signal, beyond the point of stimulation, he sent a barrage of letters to his friends and botanical correspondents. But such was his enthusiasm for the new phenomenon that he went beyond his normal circle. Among those who in June 1873 learned of his discoveries was John Burdon Sanderson, Professor Superintendent of the Brown Institution (University of London) and Professor of Practical Physiology at University College. If Sanderson's excellent scientific credentials had needed underlining for Charles, then Francis was the man to do it, for he was in 1873 a part time student at the Brown Institution where he was carrying out, under the supervision of Sanderson's protégé, E. E. Klein, a research project that contributed to the MB degree for which he was studying.

Sanderson, now in his early forties, had already earned the highest reputation as a pathologist but he was switching his attention to physiology, embarking on his pioneering studies of muscle contraction.[47] A passionate amateur botanist – he was a national expert on the cultivation of lilies – he also had some professional interest in plants, having been inspired as a student in Edinburgh by the teaching of James Hutton Balfour and having himself held for one year the post of lecturer in botany at St Mary's Hospital in London. He was a very busy man, however, with his immediate career in animal physiology mapped-out, so he was not instantly persuaded to turn to the physiology of plants, even when entreated to do so by the great Charles Darwin. What finally persuaded him was a practical demonstration of the movement of *Drosera* tentacles that Charles had specially arranged for him, together with Joseph Hooker, at Down House. Sanderson was recruited, being convinced that insectivorous plants had something to teach him, something that might help illuminate his studies of nerves and muscles. Little did he know then that, when in his old age he was to look back on his career, it would be his work on plants that would give him most satisfaction.[48]

Almost like luring his prey into a trap, Charles tempted Sanderson with not just the marvels of *Drosera* but, also, with the amazing snapping leaves of *Dionaea*. Its leaves are bi-lobed and are fringed by extensions, like the tines of a fork, that interdigitate when the lobes snap together. Such rapid closure occurs in less than one second and is triggered whenever an insect touches one or more of the three trigger hairs found on each lobe – multiple stimulation of one or more hairs is required for closure. Smaller insects can escape from between the interlocked tines, but larger ones are effectively caged. The struggles of trapped insects and any nitrogenous compounds they excrete promote the secretion of acid and digestive enzymes by the leaf. Only when the prey has been fully digested does the leaf relax and open.

Sanderson's laboratory was fitted with the most modern equipment and his own dexterity and manipulative skills were outstanding. With the help of his invaluable and but often overlooked assistant, F. J. M Page, he had modified and improved an apparatus for measuring the very small electric currents that are generated by animal nerve cells or, as he was soon to find out, by the snapping leaves of the Venus Flytrap. Carried rapidly forward by his enthusiasm for the new subject, and with some help from Francis, he was by late 1873 able to give his first report on *Dionaea* to a meeting of the British Association for the Advancement of Science (BAAS). He astonished his audience by telling them how he had been able to measure electric currents in leaves of a *plant*, the Venus Flytrap, when its trigger hairs were touched by insects. And then he amazed them further by demonstrating how, when he applied electrodes to one or other of the leaf lobes of the Flytrap, he could cause that lobe to flex as he supplied an electrical stimulus.[49] As Francis noted in the second edition of *Insectivorous Plants*, Sanderson was by 1887 confident enough to proclaim that the signal was propagated at a rate of 100 millimetres per second, 'just about the rate of propagation of the excitatory disturbance in the muscular tissue of the heart of the frog'.[50]

But what *mechanical* forces caused the leaf to close? Sanderson drew on his superior knowledge of the literature to provide the best answer that was possible at the time. He made the connection between, on the one hand, the irritability of plants and, on the other, what Dutrochet and Pfeffer had discovered about osmosis and the turgor of plant cells and, incidentally, what Charles had missed. In lectures delivered to the Royal Institution in 1882,[51] and elaborated upon in the *Philosophical Transactions of the Royal Society* in 1888, Sanderson argued that movements, from the closure of *Mimosa* leaves to the contraction of stamens of *Centaurea*, had in common a rapid loss of turgor and, therefore, of cell volume:

> The protoplasm suddenly [losing] its water-absorbing power, so that the elastic force of the envelope at once comes into play and squeezes out the cell contents.[52]

Sanderson was still faced with the problem of how an electrical stimulus was converted into a mechanical one (he sidestepped the complementary question of how the leaf hair turns a mechanical stimulus into an electrical one). It was too early in the history of electrophysiology for that first question to be answered in full but Sanderson was clearly thinking along the right lines. He drew an analogy between 'continually tightened springs' and [unstimulated] plant cell membranes. Today it is recognized that differences across membranes, in both the concentration of soluble chemicals (including electrically neutral solutes, such as sugars, and positively or negatively charged ions into which salts dissociate in aqueous solution) and in electrical potential, set up by energy-consuming pumps located in the membranes, are the key to understanding the role of turgor in plant movements. The combined *electrochemical* difference established between the inside and the outside of the cell is a form of potential or useful energy. It is located *across* the cell membrane rather than *in* it as Sanderson thought. What an electrical stimulus does is to alter the difference in electrical potential across the membrane, resulting in a rapid redistribution of ions, which in turn generates an osmotically driven flow of water.

By 1888, Sanderson had finally concluded, though he couldn't prove it, that in the leaves of the Fly trap, as in other instances of rapid movement in plants, a rapid change in the turgescence of *critically located* cells was responsible for changes in the relative position of tissues.

> By turgescence we understand the power which living protoplasm possesses of retaining water. In the case of cells which are excitable, the immediate effect of excitation is suddenly to diminish this power, and thereby to produce a diminution of volume of the cells which is equal to that of the water (probably holding diffusible bodies in solution) which is discharged into the intercellular spaces.[53]

Although today it is clear that solutes and water are lost to the internal, non-living spaces of the leaf as cell turgor collapses, the exact location of the critical motor cells in the leaves of *Dionaea* is still a matter of dispute.

*****

From the 1873 meeting of the BAAS onwards, animal physiologists, many of whom were doctors of medicine, were forced to give plants a new respect. Joseph Hooker observed joyfully, 'Not merely then are the phenomena of digestion in this wonderful plant [the Fly trap] like those of animals but the phenomena of contractility agree with those of animals also'.[54] Regrettably, he was missing the point. Now in his mid-fifties, he was looking backward to an age when plant functions were explained in animal terms. Such an eminent botanist should have been looking forward, rejoicing in the fact that Sanderson had offered one more reason why plants were worthy of study for their own sake.

Sanderson, the animal physiologist, had helped botany shake off one more of medicine's shackles.

7 CHARLES DARWIN, FRANCIS DARWIN AND DIFFERENCES WITH VON SACHS

Like many a young man who has enjoyed a comfortable childhood, Francis Darwin simply drifted into a career, in his case medicine. There is no record that the choice caused him any great anguish or soul searching. There were, after all, plenty of precedents in his family for entering medicine, albeit often unhappy ones, and medicine seemed a sensible choice given his general interest in biology, encouraged in childhood first by his father and later by his school. He might not have had a great passion for the subject itself, or have been driven by any religious vocation to heal the sick, but there is no evidence that he was strongly pulled in any other direction. Family and friends alike commented that, above all, Francis was a steady man, not given to whims or transient enthusiasms. He was, however, burdened with a heavy sense of duty.

On graduating in 1870, aged twenty-two, from Trinity College, Cambridge with a BA degree (first class), he moved on to St George's Hospital, where he registered to take an MB degree from the University of London. Required to carry out small pieces of research and to write a thesis towards his final degree, Francis soon found an interest in, and flair for, physiology. Perhaps he was enthused with the spirit of the age, for these were indeed exciting times for the subject in Britain. New ideas and approaches, already well established in France by men like Claude Bernard, and in Germany by Carl Ludwig and others, were at last infiltrating Britain's staid institutions. Leading the establishment of physiology as a respected discipline in British university laboratories were Michael Foster and John Burdon-Sanderson.

It was Foster who in the late 1860s had introduced at University College, London (UCL) the very first course in experimental physiology in Britain. When in 1870 Foster moved on to a chair in Cambridge, he was replaced by Sanderson. Quickly realising that he could not on his own make the laboratory a success, Sanderson strove to build around himself a group of the most able young physiologists. One he fought especially hard to recruit was E. E. Klein, whom he persuaded to move from Vienna to London.[1] It was under Klein's supervision at the Brown Institution – a part of UCL privately endowed 'for investigating, studying and … endeavouring to cure maladies, distempers, and injuries' of 'Quadrupeds or Birds useful to man' – that Francis was lucky enough to carry out his research projects.[2]

His investigations clearly went well, for three research papers resulted and his work was admired by no less an authority than Foster. An unexpectedly bright future was opening up for Francis. A career as a medical practitioner was no more or less appealing at this juncture than it ever had been, but an exciting career as a medical researcher was a different matter. Francis must sometimes have reflected, with justification, that perhaps he had found his niche in life – as a researcher in the fast expanding world of animal physiology. And of no less importance, he had fallen in love with Amy Ruck, and she with him. His life appeared calm and settled, and his future mapped out, but in the next few years everything would be turned upside down. He would marry, then lose, Amy (see Chapter 2). He would forsake medicine and animal physiology for plant physiology, and he would dedicate a large part of his life to being his father's secretary and, later, his biographer. He would emerge from the next ten years a very different person.

*****

The research in which Francis was involved gives some insight into the skills and expertise he learned and which he would later bring into a working partnership with his father. His study, of 'the Anatomy of the Sympathetic Ganglia of the [Mammalian] Bladder in their Relation to the Vascular System', published in 1874,[3] clearly drew upon Klein's histological expertise. Progress was made because Klein had found a way of detecting the nuclei of cells of nerve ganglia by staining with gold chloride or picrocarminate of ammonia. From the anatomy of bladders, Francis turned to a more general study of the relation between locally applied inflammatory irritants, vasomotor nerves, and the contraction and dilatation of blood vessels. His first results, included in both his MB thesis (1875) and a paper published in 1876,[4] were oddly anti-Darwinian where they implied that contraction of the heart might have different origins in different organisms; a neurogenic (nervous) origin in some and a myogenic (muscular) origin in others. Francis's approach and findings nevertheless attracted the attention and admiration of Michael Foster who was attempting to demonstrate in a range of animal systems – ultimately with success – that contraction was *universally* myogenic, having the common origins consistent with Darwinian theory. Foster suggested to Francis that he should undertake an investigation of the snail's heart, which he did, his publication of 1876[5] clearly showing the absence of any nerve ganglia in the heart, or any nerves leading to the heart. Francis concluded that the contraction of a snail's heart must, therefore, be myogenic – as Foster had hoped.[6]

Foster would have liked to recruit Francis to the growing ranks of animal physiologists but the odds were always stacked against such an outcome because Francis could not give priority to the subject. He could not commit himself fully to animal physiology because, as early as September 1873, Charles and his brother Ras[7] had discussed the possibility that Francis might become his father's desperately needed secretary. Charles was troubled. The enormous volume of his correspondence, not to mention the continuous demands of editing the manuscripts of new scientific papers and books, and revising the older texts, was becoming too much for him to cope with. His daughter Henrietta read proofs and helped to edit some manuscripts but this was not enough. Against all his instincts, Charles was forced to contemplate employing a paid assistant from outside the family. Rejecting this idea, his solution, which typically drew upon strengths within the family, was for the dutiful Francis to do the job while continuing his studies of the snail. Inevitably it meant Francis giving up for the foreseeable future any thoughts of a full time career in physiology or medicine. But Francis was probably was content, for he had other things on his mind. In 1874, he and his new bride, Amy, moved into a cottage in Downe village and, unless he was away researching the snail, Francis needed to walk only a quarter of a mile up the lane to his father's house where he would pass each day as his secretary.[8] And, until Amy's death in 1876, that was how his life stood.

*****

The lasting influence of the training that Francis received from Klein, and his experiences working alongside other researchers in an animal physiology laboratory, including a period in 1873 helping Sanderson to measure electric currents in plants,[9] cannot be underestimated. He learned practical, manipulative skills: how to draw, observe and record. Perhaps most important of all, he saw how purpose-built laboratory apparatus could reveal the secrets of living organisms, as in Sanderson's laboratory where physiologists were using sophisticated equipment to detect and amplify minute changes in the position, contraction, and electrical activity of nerves and muscles. The contrast with his father's method of working could not have been more striking. If Francis's character had been other than equable, and his love and respect for his father anything less than profound, he might have become deeply frustrated and restless. As it was, he was able to work productively with his father for eight years, their skills complementing each other's and any tensions held well in check.

For his part, Charles cannot have fully appreciated what Francis had given up. Although he kept himself well informed about the work of Sanderson and Foster and their latest discoveries, Charles had no experience of working in a well equipped modern laboratory, or of the way the thrill of discovery is enhanced when the researcher is part of a group or team. However, he did recognize that his son had already acquired the skills and perhaps more importantly the habit of practical investigation and, fearing that he might become bored by humdrum secretarial duties, Charles soon encouraged Francis to involve himself in small pieces of research. In the first instance he asked his son to take over some small problems whose solution would progress his own projects. And, since Charles was at this time deep in his botanical investigations, these were botanical problems. Francis had found his métier. By 1876 he was able to publish as sole author the first of a string of papers on insectivorous plants, in this case in the *Quarterly Journal of the Microscopical Society*, on 'The Process of Aggregation in the Tentacles of *Drosera rotundifolia*'.[10] Charles proudly told his good friend Hooker about Francis's progress and encouraged his son to correspond in his own right with members of the informal circle of botanical experts that he had gathered around himself over many years.

One of Francis's projects was to study the glandular hairs of teasel. He proposed that the leaf hairs had swapped their original secretory function for an absorptive one. The teasel's paired leaves formed cups around the stem into which rain and then insects fell, and its glands absorbed ammonia, Francis claimed, from the 'decaying fluid'. When the Royal Society rejected for publication the full text of Francis's paper, 'On the Protrusion of Protoplasmic Filaments from the Glandular Hairs of the Common Teasel, *Dipsacus sylvestris*', Charles was outraged – much more so than was Francis.[11] Determined that his son's work should receive the attention he believed it deserved, Charles used his influence to ensure that the paper was read to a meeting of the Royal Society on 1 March 1877 (with his friend Joseph Hooker in the chair), guaranteeing that at least a lengthy abstract would be published in the *Proceedings of the Royal Society*. He told one of his correspondents, the distinguished botanist Ferdinand Cohn, about Francis's findings, and when to Charles's great satisfaction Cohn replied that he had confirmed most of Francis's conclusions, Charles wrote triumphantly to *Nature*:

> The observations of my son Francis on the contractile filaments protruded from the glands of *Dipsacus*, offer so new and remarkable a fact in the physiology of plants, that any confirmation of them is valuable. I hope that therefore you will publish the appended letter from Prof. Cohn, of Breslau, whom every one will allow to be one of the highest authorities in Europe on such a subject. Prof. Cohn's remarks were not intended for publication, but he has kindly allowed me to lay them before your readers.[12]

Thus, Charles made sure that in his *Letter to the Editor*, an extract from Cohn's correspondence was included:

> In the meantime I am happy to congratulate Mr. Francis Darwin and yourself on account of the extraordinary discovery he has made, and the truly scientific paper in which he has elaborated it, and which has added a series of quite unexpected facts to the physiology of plants.[13]

Botany was inexorably drawing Francis into its clutches. When, following the death of his father, he moved to Cambridge in 1882, Foster tried to involve him in teaching animal physiology. Francis was enthusiastic but Foster's plans were ill-timed (see Chapter 7). Nevertheless, they remained good friends, their families happily socialized and professionally Francis benefited greatly from Foster's goodwill and generosity – not to a lecturer in animal physiology, but to one in botany.

*****

One of the ways in which Francis was able to help his father in those first years back at Downe was in the production of a second edition of *The Movement and Habits of Climbing Plants*, published by John Murray in 1875. In the ten years since the first edition, further experiments had been conducted to amplify and clarify the original thesis, and some of this new material was included. However, with a fresh view on the subject, and with insights and questions provided by Francis, Charles wisely decided to limit the scope of the revision and to take forward in a different, more focussed direction, his interest in movement. Believing that circumnutation could explain not just the twining of climbing stems, leaves and tendrils, but also the way in which roots sought gravity, shoots sought the light, and even the sleep movements of leaves, Charles, with the help Francis, felt able to embark on a new, exhaustive, and wide ranging series of experiments. Their outcome was a new book, *The Power of Movement in Plants* (1880), widely regarded as Charles's most important botanical publication because of its originality and lasting impact on the science. Francis's part in its production has often been overlooked, which is not surprising, given that the book's title page reads, 'By Charles Darwin Assisted by Francis Darwin', but this is a great shame because the originality of the materials studied, the book's balance of descriptive observations versus quantified experimental results, and its awareness of the current literature, bear the hallmarks of input from a younger mind.

The judgement of history has been generous to the book but at the time of its publication it was viewed much less favourably, largely because of the antipathy of one man, Julius von Sachs.[14] A string of achievements in different areas of the subject had firmly established Sachs by 1880 as the world's pre-eminent plant physiologist. He had pioneered methods of water culture that made possible rapid advances in the study of plant mineral nutrition (ultimately leading to methods for growing crops hydroponically). He had built on the work of Ingen-Housz and de Saussure (see Chapter 4) and shown that the primary formation of starch occurs only in association with chlorophyll, which is itself confined to tiny discrete bodies – later called chloroplasts – within what had previously been thought of as *uniformly* green cells. He showed that starch formation has an absolute requirement for light and carbon dioxide. (This could be dramatically demonstrated by showing that when parts of the leaf are shaded, subsequent staining with iodine clearly reveals that starch formation is restricted to the illuminated areas). He demonstrated that starch produced through photosynthesis is later broken down into sugars before being translocated out of the leaf to either growing points or storage tissues; Sachs clearly distinguished for the first time long-lived storage compounds, such as starch and oils, from those shorter-lived compounds, such as sugars and proteins, that are more directly involved in growth. And he had led the way in studying the effects of temperature and light on processes such as seed germination, flower opening, and transpiration. In particular, he recognized the peculiar sensitivity of plants to blue light, anticipating by many years final proof that the blue part of the solar spectrum governs phototropic responses.[15] If more were needed to spread his influence, Sachs provided it through both his much admired textbook, *Lehrbuch der Botanik* (1868), which was translated into English and several other languages, and through the stream of botanists he trained in his laboratory. Selected from the most talented young men of Europe and North America, they disseminated his philosophy of teaching and learning worldwide, passing on to their own students Sachs's interpretation of plant functions. The best of the Würtzburg graduates, like Pfeffer, were destined to take over the leadership of plant physiology.

Sachs more than any man has a justifiable claim to be called the father of plant physiology but, like all men, he had his weaknesses and prejudices. He may have begrudged the Darwins their wealth and ease. Sachs's early life had been difficult for his father had been a lowly engraver and both his parents had died before he was eighteen years old.[16] His objections to the Darwins were probably nationalistic as well as scientific. All the sciences were becoming professionalized in the mid-nineteenth century. They were proudly led by a relatively small, elite group of German professors. Thus, young chemists from Britain, as from elsewhere, were flocking to study under Liebig in Giessen, while botanists were beating a path to the doors of Sachs in Würtzburg and of Anton de Bary in Strassburg. As Charles Dodgson, mathematics tutor at Christ Church College, Oxford – *aka* Lewis Carroll – grumbled:

> Now-a-days, all that is good comes from the German. Ask our men of science: they he will tell you that any German book must needs surpass an English one. Aye, and even an English book, worth naught in its native dress, shall become, when rendered into German, a valuable contribution to science ... No learned man doth now talk, or even so much as cough, save only in German. The time has been, I doubt not, when an honest English, 'Hem!' was held enough, both to clear the voice and rouse the attention of the company, but now-a-days no man of science, that setteth any store by his good name, will cough otherwise than thus, Ach! Euch! Auch![17]

For Sachs, German laboratories were the model for the future: the past was exemplified by Charles's study at Down House.

The opening words of *The Power of Movement in Plants* set out succinctly the major postulate the Darwins were attempting to prove; it was that plants move according to well-set patterns.

> The chief object ... is to describe and connect together several large classes of movement, common to almost all plants. The most widely prevalent movement is essentially of the same nature as that of the stem of a climbing plant, which bends successively to all points of the compass, so that the tip revolves. This movement has been called by Sachs 'revolving nutation'; but we have found it more convenient to use the terms *circumnutation* and *circumnutate*.[18]

This is followed by a description of the basic method by which the irregular ellipses or ovals described by shoots, roots and leaves were recorded – over one hundred of the several hundred that the Darwins traced are reproduced in *The Power of Movement* (most are similar to those Charles had published in *Climbing Plants*.

The Introduction to *The Power of Movement*, like the rest of the book, is not well referenced, but it does draw upon published sources more often than did Charles's earlier books. In those, he had typically cited only the discoveries and opinions of his limited circle of correspondents – Charles was the master of what today is known as the unpublished personal communication, or *pers. comm.* – seldom giving references to the wider body of published botanical literature. The fact that references to the wider literature are more common in *The Power of Movement* probably bears testimony to Francis's greater knowledge of recent papers and the people writing those papers. Not only was he familiar with the ways of a modern science, having been exposed to the practices of animal physiology, but he had made numerous personal contacts with contemporary researchers during his two visits to Sachs's laboratory in 1878 and 1879. Familiarity with recent publications would have been demanded by Sachs, a rich source being his own *Lehrbuch*, of which Francis was able to take advantage since he was able to read and speak German fluently. (In contrast, Charles could not, for even his reading of the language was notoriously laboured.) Thus, in addition to citations of Sachs's work, *The Power of Movement* mentions relevant publications by several German researchers. Frank and Wiesner are frequently cited and, with reference to publications of Pfeffer (see Chapter 6), the importance of turgor is for the first time noted. Particularly persuasive, says *The Power of Movement*, is the work of a Dutchman, Hugo de Vries, published only a year earlier (1879):

> Until recently the cause of bending movements was believed to be due to the increased growth of the side which becomes for a time convex; that this side does temporarily grow more quickly than the concave side has been well established; but Dr Vries has lately shown that such increased growth follows a previously increased state of turgescence on the convex side.

The Power of Movement in Plants reaches the simple conclusion:

> increased turgescence of the cells, together with the extensibility of their cell walls, is the primary cause of the movement of circumnutation.[19]

It was probably thanks to Francis's ability to read German that the Darwins were aware of and gave generous acknowledgement to the recent advances in the understanding of root physiology made by Theophil Ciesielski, of which there is more below. Their omission of any reference to Andrew Knight is odd, first, since his discoveries about roots, like those of Ciesielski, were praised by Sachs and, second, since Charles definitely knew of Knight's botany, admiring his work on the subject of cross-fertilization (see Chapter 5).

Knight was an English gentleman scientist in the same tradition as Charles Darwin. Mainly interested in apple breeding and similarly practical aspects of horticulture, he occasionally ventured into more physiological matters, and always with stunning ingenuity. Noting that the radicle or first root of a seedling typically grows downward, Knight attempted to determine whether this is a response to gravity (positive geotropism=gravitropism) or to moisture in the soil (positive hydrotropism).

> Having a strong rill [stream] of water passing through my garden, I constructed a small wheel similar to those used for grinding corn ... Round the circumference of the latter ... numerous seeds of the garden bean, which had been soaked in water to produce their greatest degree of expansion, were bound... The radicles were made to point in every direction ... The water being then admitted, the [vertical] wheels performed something more than 150 revolutions per minute ... In a few days the seeds began to germinate ... the radicles, in whatever direction they were protruded from the position of the seed, turned their points outwards from the circumference of the wheel, and in their subsequent growth receded nearly at right angles from its axis. The germens [shoots], on the contrary, took the opposite direction.[20]

This orientation, he argued, must be caused by the centrifugal force developed, thus showing that such a force can act like gravity in directing the orientation of root and shoot growth. Sachs later copied the principle of Knight's wheel and designed the klinostat, an ingenious but simple piece of equipment used ever since by physiologists to demonstrate the effects of gravity.

It was the radicle that occupied most of the Darwins' experimental time and a description of its properties occupies most pages of *The Power of Movement*. They concentrated their experiments on one species, *Brassica oleracea* (the cabbage) but, following Charles's well established habit, when they had reached their primary conclusions on the basis of extensive investigations of one organism, these were then tested less extensively on a wide range of species, wild and domesticated, indigenous and exotic.

The method used with cabbage and most species was to fix the seed to a small zinc plate, the 0.05 inch-long radicle pointing upwards. A fine glass filament was then attached near the base of the radicle, close to the seed coat, in order to magnify any movements that might follow. Roughly circular movements were observed and these were traced during the next sixty hours as the radicle extended to a length of 0.11 inches, soon turning downwards. With some other species, such as horse chestnut and bean, an alternative method was used in which the radicle was allowed to grow across an inclined smoked glass plate. The root tip disturbed the soot as it extended, leaving a snail-like trail as evidence of its twisting and turning. The conclusions were clear: the tip alone, for a length of only 2/100 to 3/100 of an inch, was acted upon by gravity. Even the basal part of the radicle, to which the filament was attached, circumnutated; the tip of the radicle, to which it was too difficult to attach a filament, described far larger ellipses.

The Darwins likened the radicle to a burrowing animal such as a mole, which in attempting to burrow down through the soil continually rotates its head to find the most suitable pathway. Thus far their conclusions were new but not potentially controversial. What challenged Sachs and his students and followers – including almost all the leading plant physiologists of the day – was the Darwins' proposal that gravity is sensed at the tip of the root while the growth response occurs in a zone of curvature some way *behind* the tip. Thus, as in the tentacles of *Drosera* or the leaves of *Dionaea* (see Chapter 6), there must be a signal or message, they thought, that passes from a receptor to a distant responder or responsive zone.

The Darwins were not the first to discover the ability of the root tip to sense gravity, for Ciesielski, a young Pole, had beaten them to the discovery by several years. Sachs knew of Ciesielski's work (although not accepting its conclusions), as did the Darwins by the time they had published *Power of Movement*. In the Introduction to their book, they wrote, 'it is the tip, as stated by Ciesielski, though denied by others, which is sensitive to the attraction of gravity ....'[21] Ciesielski had been a student at Breslau where his PhD thesis, 'Studies on the Downward Curvature of Roots', appeared in 1871. In the next year he published his findings in a new journal, *Beitrage zur Biologie der Pflanzen*. Here he reported that curvature of roots under the influence of gravity takes place in the region of extension growth and results from greater extension of cells on the upper side than on the lower. Sensitivity is lost, he reported, if roots are decapitated, but can be regained if a new tip is regenerated.[22]

For some reason, Sachs, or rather his assistant Emil Detlefsen, was unable to reproduce Ciesielski's results. The Darwins suggested that it was because Sachs, in preparing his radicles, had not cut them strictly at right angles to the main axis of the root – a somewhat implausible suggestion that must have profoundly irritated Sachs, who prided himself on his exemplary laboratory techniques. It is always difficult to determine in retrospect why experiments with living organisms sometimes fail to produce the 'expected' or commonly observed effect. It seems likely in this case, however, that in a quest for rapid responses Detlefsen experimented on radicles at too high a temperature. He had failed to learn from the Darwins, who had included in their studies a lengthy examination of the effects of temperature on the sensitivity to gravity of the root tip of the garden pea (*Pisum sativum*), finding that most peas lost their sensitivity to gravity at temperatures in excess of 70° fahrenheit.

Minor insults continued to be traded. In his turn, Detlefsen criticized the Darwins for having used dry and injured radicles. He had found that eleven out of twelve decapitated radicles exhibited curvature when laid horizontally; four curved sideways, and one upward, but six downward, which he and Sachs took as evidence that the tip was not necessary for gravity sensing. His results were published first in the journal of the Würtzburg Institute (1882) and soon after were quoted by Sachs in his book, *Vorlesungen uber Pflanzenphysiologie* (1882), together with the withering comment:

> In such experiments with roots not only is great precaution necessary, but also the experience of years and extensive knowledge of vegetable physiology, to avoid falling into errors, as did Charles Darwin and his son Francis, who, on the basis of experiments which were unskilfully made and improperly explained, came to the conclusion, as wonderful as it was sensational, that the growing point of the root, like the brain of an animal, dominated the various movements of the root.[23]

Elsewhere Sachs wrote:

> Personal acquaintances often have their good side. I first became aware of the whole wretchedness of Darwin's activities when Francis Darwin studied here in 1878 and 79; I had the opportunity to look behind the scenes and when the miserable book 'On Movements' appeared, I realised that here we are dealing with literary rascals.[24]

Sachs's scorn was not unexpected, but was perhaps all the more hurtful to the Darwins since Francis had chosen to go to Würtzburg, in humility, to learn from Sachs: an act which was in itself a recognition of Sachs's eminence. Moreover, the Darwins had never tried to usurp Sachs's position as the ultimate authority in matters of plant physiology. The second time Francis visited Würtzburg, he had gone in the hope that he and Sachs could discuss constructively and maturely, an area of research in which both men had new, although contradictory, results. Francis recollected with a trace of sadness:

> Sachs was most kind and helpful [during the first visit], and under his direction I contributed a small paper to his *Arbeiten*. ... I am sorry to think my relationship with Sachs came to an unhappy ending. I published what seemed to me a harmless paper, in which I criticised some of his researches. I wrote to him on the subject but received no answer. Partly on account of his silence and partly to pay a visit to a friend, I travelled to Würtzburg. I found Sachs in the Botanic Garden; he seemed to wish to avoid me, but

> I went up to him and asked him why he was angry with me. He replied: 'The reason is very simple; you know nothing of Botany and you dare to criticise a man like me'. I had no opportunity of replying, for at that moment one of his co-professors addressed him, asking if he could spare a moment. 'Very willingly, Herr Professor,' said Sachs, and walked off without a word to me. And that was the last I saw of the great botanist. I was undoubtedly stupid, but I do not think he showed to great advantage in the affair.[25]

Although Sachs praised Ciesielski's work, he dismissed his conclusions, as he had those of the Darwins. Gravity was sensed, he proclaimed, *throughout* the growing region of the root. His authority ensured that, at least for the next few years, his view was the accepted orthodoxy. Paradoxically, although time was his enemy on this specific point of plant physiology, it was to prove his ally where he and the Darwins diverged in their general approach to botanical research, in that, as will be seen later, it was his style which was to prevail.

If the root tip described circles, searching out the path of least resistance as it extended downward through the soil, what happened when it reached a rock or a stone, or an artificial barrier such as a glass plate which the Darwins might put in its way? The answer was that the root sensed the barrier and, at least temporarily, changed its direction of growth, in the process, flattening itself against the obstruction. Charles and Francis hypothesized that sensitivity to touch can override sensitivity to gravity – thigmotropism can overcome gravitropism. Charles had previously noted sensitivity to touch in the way that climbing leaves and tendrils wrapped themselves around supports, such as his own walking stick (see Chapter 6). Here in the root was an analogous movement, or tropism.

In both climbing stems and roots the *observable* response is slow, for it depends on a decrease in the rate of extension growth and an increase in the rate of radial growth. However, in cells at the point of contact very much faster responses can be detected with the aid of today's equipment. The earliest responses, in the electrical resistance of the plasma membrane, can be measured within seconds of the contact stimulus. There appears to be a strong correlation between the degree of longitudinal strain experienced and the extent of the response, probably because stretching the membrane affects ion transport into and out of the cell and thus the electrical potential of the plasma membrane. (Similar stretch sensitivity explains how a mechanical stimulus can be converted into an electrical signal, for example, in the basal cells of the hairs of a Fly-trap's leaf; see Chapter 6.) Among ion movements, those of calcium ions (Ca^{2+}) are critical because these ions can regulate metabolism through their effects on the activity of enzymic proteins; very small changes in calcium ion concentrations can thus have major effects on function.[26]

The Darwins' admiration for the root tip knew no limits: in a comment which might have come from Erasmus's pen, and which they might have regretted since it provided ready ammunition for Sachs, they wrote:

> We believe that there is no structure in plants more wonderful ... It is hardly an exaggeration to say that the tip of the radicle ... acts like the brain of one of the lower animals.[27]

It was not only the direction-finding processes of the seedling root that Charles and Francis investigated. They also explained how, in plants such as beans, circumnutation in combination with the arching of the young hypocotyl (the region of the stem below the two cotyledons or seedling leaves) helps the young plant to burst from the soil while protecting its delicate shoot apex. The general theme of circumnutation took them forward to explore the circling movements of the cotyledons themselves and those of shoot apices, the latter investigations extending several aspects of Charles's earlier work on climbing plants.

In a related area of physiology, the sleep movements of leaves, the Darwins took a well known phenomenon and provided a set of observations that in their scope, precision and description went far beyond any made before. Thus, father and son did for sleep movements what Charles had previously done for subjects such as floral function in orchids or climbing mechanisms – they laid firm foundations upon which others could build.

The phenomenon of 'sleep' in plants – named nyctitropism by the Darwins but more correctly called nyctinastsy wherever the direction of movement is independent of the direction of the stimulus – had been noted as long ago as the first century AD by Pliny the Elder and had also been the subject of a well known essay by Linnaeus, *Somnus Plantarum*. In many plants, especially legumes, the leaflets of compound leaves fold together at night. In some species they gather upwards, in others downwards. Pointing to their own observations that leaves whose leaflets were physically prevented from closing normally at night were often killed at lower temperatures, while leaflets allowed to close survived similar temperatures, the Darwins proposed that a plant protects itself from 'radiation at night', i.e. chill damage, by leaf closure. It is now thought that in many species the benefit of leaf closure might rather be that of conserving water. If opposing leaf surfaces are brought together at night, when photosynthesis is impossible and there is no reason to maximize the leaf area displayed, then transpiration is minimized. Charles and Francis's list of plants exhibiting nyctitropism was long; the list of those to which they devoted special attention was shorter, and included some familiar names, *Oxalis*, *Mimosa* and the plant so loved by Erasmus Darwin, *Desmodium gyrans* (see Figure 3.1).

The physiological mechanism of leaflet movement in nyctinasty is now well understood and is comparable to that powering other rapid leaf movements, such as the snapping closed of the Venus Flytrap or the rapid collapse of the leaf of the touch-sensitive *Mimosa pudica*.[28] Movements result from differential changes in the turgor of cells located on opposite sides of the pulvinus, a specialized structure at the base of the petiole (leaf stalk). While leaflets are opening, the ventral motor cells accumulate potassium (K^{+}) and chloride (Cl^{-}) ions, causing the osmotic uptake of water and turgor-driven swelling, and simultaneously, the reverse happens in the dorsal motor cells which shrink as K^{+} and Cl^{-} ions are released. When the leaflets close, fluxes of ions and turgor changes are in the opposite direction (Figure 7.1). In young plants a rhythm of sleep movements is induced by small changes in the solar spectrum that occur naturally through the course of each day. At dawn the spectrum is relatively rich in blue light, stimulating opening, whereas at dusk it is relatively rich in red light, stimulating closing. The rhythm becomes entrained, persisting for many cycles in the absence of any external stimulus.

Scientific knowledge typically progresses from day to day by small steps but occasionally there is a giant stride forward, a quantum leap. Those arising from a new theory or hypothesis attract most attention in hindsight, possibly because they celebrate the power of human intellect. Frequently, however, major advances are reliant not upon one man or woman but upon a new instrument, a novel chemical, or, in the case of biology, the bringing into the laboratory of an organism whose genetics, physiology, or biochemistry is exceptionally easy to understand and later to manipulate. Thus, in the 1930s and for several decades, the common fruit fly, *Drosophila melanogaster*, became the standard organism – indeed almost the only organism – used worldwide for studies of genetic inheritance. In much the same period, pigeon breast muscle became the common currency of biochemists studying respiratory metabolism. So it was that, thanks to the Darwins, the coleoptile of *Avena sativa* (oat) was for decades the

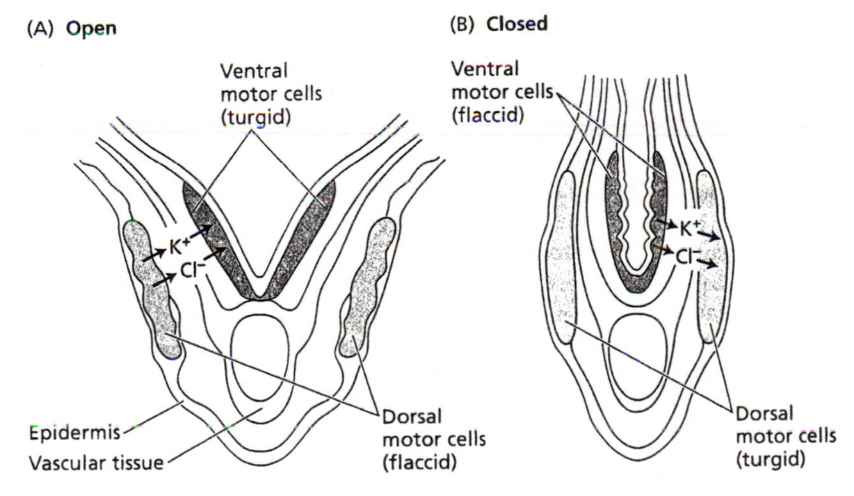

Figure 7.1: Fluxes of potassium (K^{+}) and chloride (Cl^{-}) ions between the dorsal and ventral motor cells in the pulvinus of *Albizia* (a close relative of *Mimosa*). Blue light causes opening (A); red light followed by darkness causes closure (B). Reproduced by permission of Sinauer Associates, US, from L. Taiz and E. Zeiger, *Plant Physiology*, 4th edn (Sunderland, MA: Sinauer Inc., 2006).

first choice material of plant physiologists studying the identity and properties of natural and synthetic plant hormones. In each case, it was if the properties of one organism, fruit fly, pigeon, or oat, was a key opening the door to a room of previously unreachable treasures.

'Coleoptile' is a name given to the single cotyledon or seedling leaf of grasses. (The grass family, the Gramineae, belong to the Monocotyledones; flowering plants with two cotyledons belong to the Dicotyledones.) The Darwins noted that the cotyledon of some grasses is sheath like and contains very little starch, suggesting to them that its function was not to fix carbon dioxide but rather to surround and protect the delicate first true leaf as it emerges from the seed upwards through potentially abrasive soil and compacted vegetation. They noticed too that while coleoptile tips circumnutated their dominant feature was bending towards light; coleoptiles were positively heliotropic (= positively phototropic), and this greatly interested them. They had found that seedlings in general were especially sensitive to light and decided first to look more closely at the behaviour of coleoptiles of an ornamental grass, *Phalaris canariensis* (Reed Canary Grass). Only after some time did they switch their attention to the more amenable *Avena*.

When seedlings, germinated in darkness, were either enclosed in tin-boxes to half of which light was admitted through a small hole in one side, or the tips of their coleoptiles were covered in thin glass tubes that were either entirely blackened or blackened for all but one narrow strip, the Darwins found that those exposed to light bent towards it, while those from which light was excluded remained unbent. Curvature did not occur in unilaterally illuminated coleoptiles if their tips (3–5 mm) had either been excised or covered with blackened tin foil (Figure 7.2). The conclusion was clear. The tip of the coleoptile is sensitive to light and responds by transmitting an 'influence' to a more basal region, 5 mm or more behind the tip, where bending occurs. Charles and Francis had reached a comparable conclusion previously when examining the responses of radicles to gravity; they had found, again, a *mobile* influence. Coleoptiles were

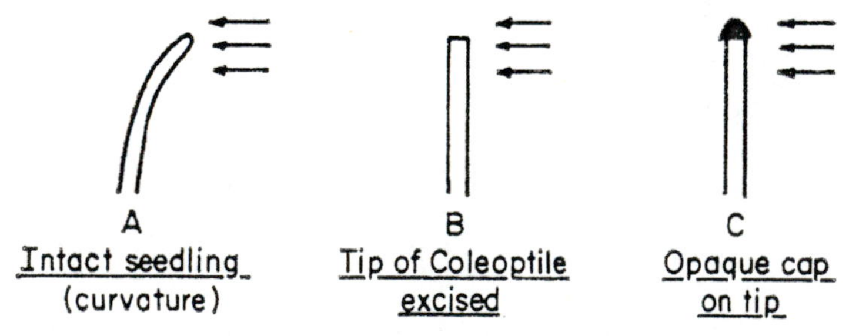

Figure 7.2: The response of an *Avena* coleoptile to unilateral illumination represented by three arrows. Reproduced by permission of Elsevier, Oxford, from P. F. Wareing and I. D. J. Phillips, *The Control of Growth and Differentiation in Plants* (Oxford: Pergamon Press, 1970).

better subjects than radicles, however, for they provided a means of quantifying the influence. The Darwins worked out that the degree of curvature, which they could measure, increased with either the duration or the brightness (intensity) of lateral illumination, which could also be measured.

The next logical step eluded them, however, because they did not state explicitly that curvature was proportional to dose, where dose=time of exposure X light intensity. Nevertheless, they had provided a platform on which others could build. Later investigators would appreciate the dose-response relationship and would refine the Darwins' methods, formalizing them as the '*Avena* coleoptile bioassay'. The bioassay enabled others to isolate and identify the mobile 'influence' before discovering that this particular influence was in fact just one representative of a much larger group of chemical hormones that govern plant growth and responses to the environment (see Chapter 9).

Predictably, Sachs disagreed with the Darwins, believing that in the case of light, as for gravity, it was the directionality of the stimulus, not the difference of intensity on the two sides of the unilaterally illuminated organ, that determined the response. He believed that it was the cells of the responding region that directly perceived the stimulus. In the short term, his trenchant criticisms held sway: most in the plant physiological community agreed with him, though they generally expressed their criticisms more kindly. One such was Julius Wiesner, Professor of Botany in Vienna. Charles was moved to write to him:

> Not a few English and German naturalists might learn a useful lesson from your example; for the coarse language often used by scientific men towards each other does no good, and only degrades science.[29]

The German plant physiologist who best appreciated the implications of a separation between the regions of perception and response was Pfeffer. Despite his allegiance to Sachs, he incorporated the Darwins' conclusions into his major text, *Pflanzenphysiologie* (1897). It was also in Pfeffer's laboratory in Leipzig that the first attempts were made to test those conclusions critically. In 1895, Friedrich Czapek inserted the flexible roots of lupin into close-fitting angled tubes of glass, each called a 'boot' because of its shape. The boots were then orientated either so that the tip or 'toe' was in a horizontal position with the upper part remaining vertical, or were held with the tip vertical and the remainder horizontal. The results vindicated the Darwins; curvatures developed, or failed to develop, according to the orientation of the tip, not that of the growing zone. More corroborative evidence was found in 1899 by Francis Darwin when he carried out a comparable experiment examining the geotropic (and phototropic) responses of immobilized coleoptiles of the grasses, *Setaria* and *Sorghum*.[30]

Another student of Pfeffer's, Wilhelm Rothert, repeated the Darwins' work on phototropism in coleoptiles. He confirmed that the tip was indeed much the most light-sensitive region of the coleoptile but he did report some sensitivity in more basal regions. About the same time, Piccard, using bean roots on a modified Knight's Wheel, reported that the tip of the radicle was, as the Darwins had found, by far, the part most sensitive to gravity, although there was some limited sensitivity in the responding zone.[31] Here then was some small consolation for Sachs – there was some sensitivity in basal regions. However, by the time he died in 1897 botanical opinion had turned in favour of the Darwins.

The dispute between Sachs and the Darwins is interesting in itself, but it has a much wider and more fundamental significance for the subject of botany. The personal struggle mirrored what was going on in the wider world – a struggle between the old and the new science that has been characterized as 'laboratory science versus country-house experiments'.[32] Laboratory science, which its advocates dubbed the 'New Botany', would be the winner. And, remarkably, one of the leading proponents of New Botany would be none other than Francis Darwin.

It is easy to dismiss Sachs as proud, arrogant and authoritarian. He was at times all of these but in his personal dealings he was, most of the time, very different. Even Francis, on his first visit to Würtzburg, had found Sachs 'most kind and helpful'. Similarly, other aspiring young British botanists who visited Würtzburg, Frederick Bower, Dukinfield Scott, and Sydney Vines, all happily acknowledged the professor's support and brilliant teaching.[33] None of these is more significant than Vines, for it was he who in 1877 began organising in Cambridge teaching and research modelled on what he had found in Würtzburg. Vines's laboratory was to pioneer the teaching of New Botany in Britain and in it Francis would soon find a home. In 1925, looking back almost half a century, Vines recalled:

> Considering that the greater part of my botanical lore had been derived from his immortal Lehrbuch ... I had no hesitation in deciding that I must go to Julius Sachs in Würtzburg, who was then at the zenith of his fame, and whose laboratory was renowned for its physiological work ... [I] was most kindly received by the Professor ... who showed me round the Institute and allotted me a table in the laboratory.
>
> I saw a good deal of the Professor in those early days, and the acquaintanceship developed into an intimacy which continued unbroken until his death in 1897.
>
> He was then engaged upon his experiments ... on the rate of the transpiration current, in which I occasionally helped him and had the opportunity of observing the precision and skill with which he devised them and carried them out.[34]

A picture is painted of a genial man who encouraged and was involved with his students. Sachs might have been the great champion of laboratory science and proof by experiment, and the man who developed the klinostat and the auxanometer (an apparatus for continuously recording the extension of shoot tips), but he did not allow his students to neglect their basic botanical and manipulative skills. His students were required to grow their own plants. One of his favourite sayings was 'What one has not drawn, one has not seen' (a dictum equally endorsed by his more famous contemporary, John Ruskin). He expected his students to match his own dexterity. One feat of which he was inordinately proud was that of 'cutting a single fresh and unembedded ovule into sections [while holding it] between finger and thumb, and spreading out the sections on a slide'.[35]

Sachs provided a professional training for his chosen students in a laboratory that was built for purpose. In his view, Charles, on whom he trained his attack, was not able to conduct experiments and could not use a microscope. At first an enthusiastic reader of *On the Origin of Species*, Sachs became steadily more critical, giving Charles no credit for the efforts he made in later editions to replace information gleaned from correspondence with his own original microscopic and experimental evidence.

Sachs's prejudices would have been confirmed if he had visited Charles's workplace in Down House. The old study, which he occupied until February 1878, can still be seen. It has been lovingly recreated by English Heritage. It is cramped, packed with bookshelves, specimen storage cabinets, tables and chairs. In the corner is a small screened area where he could relieve himself without the need to leave the room. The new study to which he and Francis moved in 1878, although much larger and with fine views over the lawned gardens, resembled the old study in that it was still an integral part of a family-orientated Victorian country house.

Hales in his parsonage, Ingen-Housz and Priestley at Bowood House, Dutrochet, Knight and Charles Darwin in their country houses, plus dozens of similar men all over Europe, had contributed enormously to the advance of botany while working outside universities or research institutes, but the time of such gentlemanly enthusiasts was now effectively over. If progress were to continue, botany would need laboratories, ever more sophisticated equipment, and money. Intellectual leadership and, just as important, the administrative support of the subject, was passing into the hands of the professionals, men who would soon recognize only the contributions of their own kind – other professionals – communicated through meetings and journals over which they exercised control.

*****

On 19 April 1882, a mere eighteen months after *The Power of Movement* was published, Charles died peacefully at his home in Kent. But there was no peace for his grieving family, for in death their father could not escape the public attention he had shunned in life. William, the eldest son, took responsibility for the funeral arrangements, while Francis and George wrote to Charles's many friends and correspondents giving the sad news of his death. Only a week later, Charles was buried in Westminster Abbey, not as he had expected in the quiet churchyard beside the small flint and brick church of St Mary, Downe, next to his brother, Ras. Pressure from the Royal Society, a petition signed by Members of Parliament, and representations from Huxley, Charles's cousin, Francis Galton, and Charles's closest friends had persuaded the reluctant family that the Abbey was the proper resting place for such an illustrious figure. The Dean of Westminster was pleased to give his 'cordial acquiescence'.[36]

*****

Francis, who had set out to be a doctor and had then been enticed by animal physiology before becoming his father's secretary and a part-time plant physiologist, was free to relinquish 'country house' science and to become a professional scientist. He would devote much of the remainder of his life to sustaining his father's memory and reputation but he would also follow a career of his own. It would be in plant physiology, that exciting new discipline born and nurtured in Germany but now being transplanted to Britain. Botany, like Francis, had escaped the ties of medicine. It was young and independent. In 1878 he wrote:

> Until a man begins to work with plants, he is apt to grant them the word 'alive' in rather a meagre sense. But the more he works, the more vivid does the sense of their vitality become. The plant physiologist has much to learn from the worker who confined himself to animals. Possibly, however, the process may be partly reversed – it may be that from the study of plant-physiology we can learn something about the machinery of our own lives.[37]

8 FRANCIS DARWIN, CAMBRIDGE AND PLANT PHYSIOLOGY

As the autumn of 1882 slipped into winter and the pressure of events and duties following his father's death slowly eased, Francis was forced to confront his own future. Which way should he turn? There were hints coming out of Oxford University that he would be successful if he applied for the newly vacant Chair of Botany. He confided to Nain Ruck, to whom more than anyone he could tell his innermost thoughts and worries:

> 20 Dec 1882
>
> I have been rather disturbed in my mind by the oxford people half hinting that they would like me to go in for the Botany professorship. There is no doubt that if I am to be a professed botanist I shall never have such a chance again ... I should dislike leaving Cambridge and giving up Michael Foster, though I think he could not blame me.[1]

His dilemma was that Michael Foster, head of the highly successful Cambridge School of Physiology, had dangled before him the prospect that there might be a Readership suitable for him in Cambridge. Given that neither the Chair nor the Readership were firm offers, and cautious by nature – to his family Frank was always sound and reliable rather than adventurous – he plumped for Cambridge. There he would be close to his mother, who had recently moved to the city, his siblings, and in familiar surroundings. Almost immediately, he felt he had made the wrong decision as Foster's plans were overtaken by changes to the rules for the medical examinations; these made it necessary for the lecture course to become more zoological than it had previously been, so ruling out any significant contribution from Francis. In another letter to Nain, he wrote:

> 11 Mar 1883
>
> It is rather a disappointment as I should have been appointed University Reader and so with a fixed recognised position and a good class ... Michael Foster has I think been rather stupid not foreseeing all this sooner. But he has behaved delightfully, and expresses very warmly his sorrow at what he calls bringing me here under false pretences. The whole thing is a sell and gives one a sort of feeling of being chucked out. But I am sure it is not better after all.[2]

Francis was appointed to a lectureship in Botany. He clearly thought that he had been misled by Foster, although, as the last sentence of his letter to Nain shows, he was not entirely unhappy with his new situation. So, forgetting his disappointment and forgiving Foster, his attitude was positive; in the face of continuing difficulties, he built a successful career in Cambridge in botany.

His difficulties stemmed from the fact that, in 1883, botany in Cambridge was in a woefully moribund state. The only spark of light or hope for its future was provided by Sydney Vines, a student of Huxley's and another protégé of Foster. To see how Francis fitted in, it is first necessary to know more about Vines

*****

Any coupling of the names Huxley and Darwin will forever conjure up images of that famous debate on the subject of evolution, held at the British Association for the Advancement of Science's meeting at Oxford in June 1860. In Charles Darwin's absence through illness, Thomas Henry Huxley, 'Darwin's Bulldog', took on and sensationally outwitted the Bishop of Oxford, 'Soapy Sam' Wilberforce. Huxley was one of that small group of close friends who encouraged, supported and protected Charles. Like Hooker and Asa Gray, Huxley gave Charles practical advice on plants and animals, information that often found its way into the evidence Charles presented in favour of his theory of evolution by natural selection. These same friends were at the forefront of the rational, informed debate that followed publication of *The Origin of Species* although, as in Gray's case, they did not necessarily agree with all of Charles's arguments.

The names Huxley and Darwin deserve to be linked in another way, though in this case the Darwin is Francis and the context is the history of the education of botanists. Huxley was responsible for initiating a fundamental change in the training of professional biologists. His revolution was, because of its enduring effects, of importance comparable to his advocacy of evolution by natural selection. A beneficiary of Huxley's method of teaching, Vines was to carry his philosophy to Cambridge where, in turn, it was eagerly embraced by Francis Darwin

Like many of the other leading characters in this book, both Huxley and Vines had moved into biology after having earlier committed themselves to medicine. However, their histories were very different for, whereas Vines's family possessed wealth sufficient to allow him to give up medicine after three years at Guy's Hospital, Huxley's family was not wealthy, so he could not give up the subject he disliked. With the help of a scholarship to Charing Cross Hospital, Huxley saw his training through to its completion. A frustrated botanist, who in his youth had attended John Lindley's lectures at the Chelsea Botanic Gardens,[3] he could see little chance of employment in botany. On graduation, he took a commission as an assistant surgeon in the Royal Navy. His ship, HMS *Rattlesnake*, spent the years 1847–51 surveying the coasts of south and east Australia. Huxley returned with a unique collection of plankton and a pledge to marry Henrietta, whom he had met in Sydney. Just as the voyage of the *Beagle* was the springboard for Charles's career, so that of the *Rattlesnake* was for Huxley's. He spent the next three years on study leave, arranging his material and writing the reports that soon led to him being elected a Fellow of the Royal Society. Fortune smiled on him in 1854. Just as he was wondering where his future lay and was again pondering how he could earn enough money to support himself without either medicine or the Navy, let alone bring Henrietta from Australia, a lectureship in palaeontology became available at the Government School of Mines in Jermyn Street, London.[4] He applied and was appointed. Quickly assessing the limited remit of the School and seeing how this might damage his career, he fought with characteristic energy and tactical brilliance for the School to become a general college of science with a much wider remit. Almost from the day of his appointment, influence gravitated into Huxley's hands. He was soon promoted to a Chair in Natural History and was one of two Professors who managed the School's accounts. By 1868 he had persuaded the government's Select Committee on Scientific Instruction that the [by now Royal] School of Mines should be given two new Chairs, one in Botany and the other in Mathematics. Just two years later, he was a member of the Duke of Devonshire's Commission on Scientific Instruction which recommended to Parliament, as Huxley strongly urged, that instruction in Chemistry, Physics and Natural History be transferred to new buildings in South Kensington, close to the Department of Science and Art which had since 1853 been responsible for promoting technical and science education.[5]

In the newly constituted 'Science School' at South Kensington (from which Imperial College, London, is directly descended), Huxley could, in laboratories that he had designed himself, give the practical instruction in biology which he had long regarded as the cornerstone of a proper training for biologists but which had been impossible in the cramped building at Jermyn Street. The express purpose of the Science School was to train the new teachers of science who were so desperately needed if Britain was to keep abreast of scientifically advanced countries such as Germany. In addition to the basic biology courses, which Huxley taught unaided, he mounted annually, from 1871, six-week courses that were held each summer for a small number (*c.* 45) of selected, experienced teachers. These were the courses which had the greatest impact on biological education because of the emphasis they put on practical teaching, and it is these courses that led to the connection between Vines and Francis Darwin.

To help Huxley meet the demands of a short intensive course, with its novel practical content, Huxley employed as his teaching assistants some of the best young biologists of the day, including Foster, William Thiselton-Dyer, Sydney Vines, and Frederick Orpen Bower. Inspired by Huxley, these talented young men carried his methods back to their own institutions. Time has proved the Science School was successful in meeting its primary objective of training schoolteachers, although some of its brightest graduates, such as Harry Marshall Ward, were lost to school teaching becoming instead distinguished researchers and teachers in higher education.[6]

It is hard to appreciate just how revolutionary was the idea that lectures should be supported by practical work in the laboratory. The doctrine that students should see for themselves the organisms they had read about, and be taught how to observe, draw, measure and take apart those organisms, was totally new in Britain. Conservative teachers thought it eroded the trust that their students should have in them for the new philosophy implied that books and teachers are fallible. The successful biologist, Patrick Geddes, remembered how as a student he was instructed by Huxley to examine the radula of a whelk; to his dismay he found that the mechanism was different from that described by his master. Huxley told him to look again, then, looking again himself, slapped Geddes's shoulder in delight, expostulating: ''Pon my word, you're right! Capital! I must publish this for you!' – and he did; he had the discovery published under Geddes's name in the Zoological Society's *Transactions*, explicitly as a correction of his own work. On another occasion, an anonymous instructor in physiology, seeing a drop of blood under the microscope for the first time, exclaimed with some surprise, 'Dear me! It's just like the picture in Huxley's *Physiology*'.[7] Clearly, the instructors too were novices in the practice of linking theory and observation. The value of such practical classes was immense. Charles Darwin wrote, 'It was a real stroke of genius to think of such a plan ... Lord, how I wish that I had gone through such a course'.[8]

With a grant of just £750 with which to pay himself, plus a teaching assistant and two demonstrators, and also to purchase equipment and materials, Huxley's first course, in 1871, ran in temporary accommodation and on a relatively small scale.[9] The next year it was repeated on a grander scale in the newly finished laboratories. Like all his teaching, the course reflected Huxley's belief that

> The study of living bodies is really one discipline, which is divided into Zoology and Botany simply as a matter of convenience ... [and that] the scientific Zoologist should be no more ignorant of the fundamental phenomena of vegetable life, than the scientific Botanist of those of animal existence.[10]

By 1873, Huxley's health was suffering from the enormous workload of planning and teaching that he had imposed upon himself. He was forced to take an extended holiday and entrusted the botanical instruction to Thiselton-Dyer. (The arrogant, mildly-eccentric but brilliant Dyer was destined to marry Joseph Hooker's daughter and succeed his father-in-law as director of the Royal Botanic Gardens, Kew.) Dyer introduced exciting but demanding new practical botanical exercises. Huxley was so impressed that, on his return, he asked Dyer to continue the botanical teaching. Dyer was happy to accede to the request on the condition that he, in turn, was allowed to recruit assistance. Among those he recruited in 1874 was Sydney Vines, chosen to be Demonstrator in the four hour long practical classes which followed each day's lecture. Remarkably, since it illustrates the shortage of experienced instructors, Vines was at the time an undergraduate at Christ's College, Cambridge.

Reports conflict on exactly how Vines was chosen but two things seem certain. First, he was already known to Foster whose course in animal physiology it was that had first attracted Vines to Cambridge, and, second, it was H. Newall Martin, Tutor at Christ's and friend and collaborator of Huxley, who offered Vines the post at South Kensington.[11] Whatever happened, Vines proved his worth in that first year, and in 1875 and 1876 he was rewarded with the extra responsibility of being Dyer's assistant. With little previous experience of botany when he started in 1874, Vines was now both a student and a teacher, constantly striving to keep one step ahead of the class, for example in the interpretation of microscopic sections of plant material which he, like most of his students, had never before seen. With its strong practical orientation, the novel course devised by Dyer fired Vines's enthusiasm. Given the well documented, Svengali-like, hold that Dyer exercised over the early career of another eminent botanist, Marshall Ward,[12] it was almost certainly the manipulative Dyer who convinced Vines, if he needed persuading, that his future lay in plant physiology rather than animal physiology.

The brilliant Vines graduated from Cambridge in 1875 with a first-class honours degree (having already won a first-class degree from the University of London, while at Guy's Hospital). In 1876 he was elected a Fellow and Assistant Tutor at Christ's College, Cambridge.[13] This was a pivotal moment in the history of British botany but, sadly, there is no record of how Christ's arrived at such an enlightened and far reaching decision.

Vines's teaching in that first year in Cambridge, 1876, was restricted to lectures given in College, where the only means of illustration available to him was drawing on a small blackboard. Painfully aware that his practical knowledge was limited to his experiences at South Kensington, he arranged to go early in 1877 to Würtzburg to learn from the great Sachs, then at the height of his fame (see Chapter 7). In taking what was soon to become a well-trodden path for young botanists, Vines had one great advantage over most of his peers, which was his mastery of German. His early schooldays had been spent in Germany, in consequence of the travels of his parents. He returned from Würtzburg for the autumn term of 1877 brim-full of ideas and enthusiasm but his desire for innovation may have come to nothing if it had not been for Foster's intervention. Perhaps remembering his own struggles to launch a practical course when he had first arrived in Cambridge, Foster generously provided the laboratory space that Vines needed for his class. With microscopes and other vital equipment bought from his own pocket, and a small group of advanced students, Vines was thus able to launch the New Botany in Britain.

Where innovation in teaching is concerned, too much credit has sometimes been given to the German example and not enough to Vines and the other South Kensington teachers for, in his *Reminiscences* of Würtzburg in 1877, Vines wrote in 1925:

> Clearly teaching was altogether secondary to research. There was none of that laborious and elaborate practical instruction for all the students which we in England have held to be essential and which lays such a heavy tax upon the time and energies of our teachers ... I could not help feeling a certain satisfaction that here in Germany there was nothing to approach our pioneer courses at South Kensington as far as practical teaching went.[14]

Although:

> The chief characteristic of German university life, as I saw it, was the dominance of research over mere learning ... To secure one's degree original work was essential; success depended more on the merit of the research-work than on the acquisition of knowledge.[15]

Vines faced an uphill struggle to establish both his new style of teaching and, also, physiological research. The major obstacle to progress was the incumbent Professor of Botany, Charles Babington. Now in his late sixties, he was unwell, often absent, and, worse still, unsympathetic to new ways. It was not until 1881 that he grudgingly made some space for Vines in the University Herbarium where up to a dozen students could do practical work. In spite of his difficulties, Vines managed to attract students of the highest calibre to his classes. Included among those who were to become eminent botanists in their own right were Bower and Dukinfield Scott (later the Honorary Keeper of the Jodrell Laboratory at Kew) and, also, Marshall Ward and Walter Gardiner (the latter discovered plasmodesmata, the cytoplasmic strands that connect adjacent plant cells across their juxtaposed cell walls) both of whom were to join the Cambridge staff in the next few years. Vines returned to Germany in the summer of 1789, this time to broaden his experience by studying with the great mycologist Anton de Bary in Strassburg. He went again to Strassburg in January 1880, moving on in April to Würtzburg where he continued his own education, learning from Kunkel new techniques for studying proteins. He clearly had a plan in mind because, in what little time in Cambridge was free from teaching, he then embarked on a programme of personal research directed at understanding the deposition and utilization of nitrogenous food reserves in seeds.[16]

Babington had given Vines responsibility for teaching 'Physiological Botany', reserving 'Systematic Botany' for himself. In the absence of available textbooks, Vines provided them by translating and revising the *Lehrbuchs* of Prantl, 1880, and of Sachs, 1882 (this was the second English version of Sachs's great book, Bennett and Thiselton-Dyer had made the first by translating the third edition of Sachs's book). Recognizing the unpopularity of Babington's teaching, Vines selflessly took it upon himself to broaden his own teaching, giving courses of lectures not just in physiology but in anatomy and the classification of cryptogams. It was not until 1883 that Vines was given some relief and encouragement when, at last, serious help arrived. In that year, Gardiner was made University Demonstrator and Francis Darwin was appointed Lecturer, largely taking over the teaching of physiology to advanced students. At last, Vines's achievements were given recognition within the University for he was appointed Reader (a new post had to be created for Vines because the Chair was still occupied).[17]

Vines remained in Cambridge for another five years until, in 1888, and already careworn, he accepted the offer of a Chair. It was not at Cambridge, where Babington still hung on, but at its great rival, Oxford University. However, from those last years in Cambridge at least two of his major achievements deserve mention. First he launched, with Isaac Bayley Balfour and Will Farlow, the *Annals of Botany* (1887), a sorely-needed English-language journal whose purpose was to provide an outlet for the growing numbers of physiological studies deserving publication.[18] Second, he wrote *Lectures on the Physiology of Plants* (1886), the first English language textbook of modern botany that had not been translated from a German original. This was a milestone, not just because it was an original work emanating from Britain, so re-establishing after a hundred years the credentials of British botany and botanists, but also because it marked the start of a period during which the leadership of botany would inexorably pass from Germany to Britain.

*****

Francis's preference for Cambridge over Oxford had much to do with his love of family. In the late summer of 1882 his widowed mother, Emma, had taken one of the most courageous decisions of her life; henceforward she would spend her winters not in Downe but in Cambridge. Two of her sons, George, a mathematics fellow at Trinity College, and Horace, founder of the Cambridge Instrument Company, already lived there, and a move to Cambridge would give Francis much more freedom to indulge his scientific interests than if she remained at Downe. Her plan was that each year she and her unmarried daughter, Bessy, would winter in Cambridge but return to Kent for the summer. She rented a house for the first winter but in the spring of 1883 she purchased The Grove, in Huntingdon Road, near the edge of the city. The house was selected, in part, because its extensive grounds enabled the recently married Horace to build his own home to the eastern side of her house. In the event, it also allowed Francis to build a house, Wychfield, on the western side. With George living close to the city centre, Emma had her family gathered around her, so she could enjoy their support through the winters of her widowhood. She returned to Down House each summer, as she had planned, until her death in 1896.[19]

Francis was in need of a home of his own because he was about to be married again. During his father's last years, Francis had been a frequent visitor to Cambridge. It is not clear when he met his second wife, Ellen Crofts, whether it was before Charles died or only afterwards and, therefore, what part his new romance played in Emma's decision to buy The Grove. It is possible that Francis and Ellen may have been brought together through Ellen's cousin, Wordsworth Donisthorpe, an undergraduate at Trinity College at the same time as Francis, and later a business partner of her brother, William.[20] Certainly, Francis shared an interest in photography with Donisthorpe (who, together with their mutual friend, Etienne-Jules Marey, was a pioneer of motion pictures). It is most likely, however, that Francis met Ellen at one of the many social functions organized by George or Horace.

Cambridge society was small and close-knit. Ellen was a also a cousin of Henry Sidgwick who, apart from being a friend of George Darwin, was a renowned campaigner for women's education. Clearly infected by the same spirit, Ellen had been at age nineteen one of the first students to be admitted to Newnham College. Four years later, she was appointed one of its first lecturers – in History and English Literature. It was a loss to the fledgling college when Ellen, aged twenty-seven, married Francis Darwin, aged thirty-five, on 13 September 1883. Organized by the bride's older sister, Henrietta, who was married to an Oxford academic, Paul Willert, their wedding was held in the small church of St Andrew's in Headington, then a village on the edge of Oxford.[21] On 30 March,1886, Ellen safely gave birth to a daughter, Frances.[22] Happily married, with a new family and soon a new home, Francis's career was about to enter upon its most productive period, one lasting roughly twenty years and ended only by the illness and death of Ellen.

*****

Academically, Francis's immediate task on taking up his Cambridge lectureship in 1883 was to support Vines's teaching and in this he went far beyond what was required or expected, as will be described later. Meanwhile, although teaching made huge demands on his time, he managed to re-establish that momentum in research which he had enjoyed while working with his father and which, once lost, is so difficult for any researcher to regain. Slowly at first, and following the theme of plant movements that he had pursued with his father, he gradually extended his interests into a new area, the movements of the cells surrounding stomata in the leaf surface. It was here, as a stomatal physiologist, that he finally proved he was an exceptionally able botanist in his own right.

What happened was that between 1880 (when *The Power of Movement* was published) and 1908, Francis published a dozen research papers, plus the texts of several public lectures, on growth curvatures in plants.[23] He became firmly established as the authority on the subject, at least in Britain, and as a result he was invited on three separate occasions in that period to give major addresses to the BAAS. In the first of these, at Cardiff in 1891, he argued that all curvatures were examples of irritability and told his audience about the widespread rejection in the 1880s of the old view of de Candolle that heliotropic curvature in shoots was merely the differential outcome of strong light on one side and weak light on the other side leading to different rates of growth on the two sides with the inevitable result of curvature. As President of the Biology Section of the meeting, he described how, thanks largely to the work of his father – for Francis was ever modest – the accepted view was now the one first put forward by Dutrochet back in 1824, namely that curvatures were not forced on the plant by external conditions but were self-generated upon receipt of an external signal.

Curvature was also the subject of his lecture to the BAAS meeting in Glasgow in 1901. This time, he brought his audience up to date with the work of Pfeffer and Czapek, who had confined roots within 'glass boots' to study geotropism, and with his own 1899 studies of geotropism in roots of *Setaria*, already described in Chapter 7.

In his address to the 1904 meeting in Cambridge, this time as President of the new Botanical Section, he outlined and gave his warm support to the recently proposed 'Statolith Theory' of geotropism. Indeed, he was able to describe experiments of his own which supported it. A group of plant physiologists, including Haberlandt, Nemec and Noll, had in 1900 argued that cells in the root tip are able to sense gravity because the starch grains which they contain always fall to the lowest part of the cell. (These specialized cells, which occur in stems as well as roots, are now called statocytes and their starch grains are found in bundles in statoliths, which are modified amyloplasts or starch-storing organelles). With any change in the orientation of these cells, the position of the grains was altered, thus altering the polarity of the cell. Confirming these proposals, Francis had observed in his own experiments that weak centrifugal forces also caused the movement of starch grains, in a mode consistent with the Statolith Theory.[24] Somewhat bizarrely, he had found also that the sensitivity of the cell surface on which moving starch grains settled was enhanced if the tissue was kept in vibration on the prong of a tuning fork so that the starch grains danced upon the surface. (Perhaps he was remembering the time when his father had asked him to play the tuba while examining the effect of its sounds on worm activity). How the landing of the starch grains on the cell surface, or plasma membrane, was translated into a growth signal could not be explained at the time. Only recently has it been found that statoliths can affect concentration gradients of hormones and, thus, cell polarity, as explained further in Chapter 10.

Concerning Francis's research on curvatures, he published one more paper that deserves brief mention, not for its scientific merit but for what it revealed about his fairness and generosity. In 1908 he happily wrote a short paper on geotropism to be included in the *Festschrift*[25] being assembled by the colleagues of Professor Wiesner of Vienna to mark his retirement. This was the same Wiesner who had been one of the most prominent critics of elements of the *Power of Movement*, albeit a respectful one (Chapter 7).

The way in which plants regulate the loss of water from their leaves into the relatively dry air surrounding them had long ago intrigued Stephen Hales, a man greatly admired by Francis; he contributed a chapter on Hales to Oliver's book, *Makers of British Botany*, in 1913, and later caused it to be reprinted in his own, *Rustic Sounds*, in 1917. Water loss and its regulation had, of course, also occupied his great grandfather, Erasmus (see Chapter 4). It is not surprising, therefore, that it took Francis's attention too. Recognizing the subject would be ideal for capturing the imagination of his students, and using his outstanding talent for designing simple pieces of apparatus to elucidate specific problems, Francis built a simple potometer to measure water loss through transpiration (Figure 8.1).[26] It consisted of a glass reservoir of water into which a cut shoot was fixed air-tight so that, as its leaves transpired, the water lost was replaced from the reservoir. The unique feature of Francis's potometer – others had designed different models – was that air bubbles were drawn from the outside through a long capillary tube. Their rate of movement could be easily measured and, thus, the effects of different external conditions on transpiration could be compared. The simple device proved useful in both teaching and research. In 1904, with the help of his brother Horace's Cambridge Instrument Company, Francis designed a much more sophisticated electrical device which measured transpiration by recording that cooling of the leaf which happens as liquid water evaporates prior to its vapour diffusing away from the leaf.[27] (This was probably the first of what has been over the succeeding one hundred years a continuous series of ever more sophisticated devices exploiting the same physical principle.)

Francis's simple potometer could answer basic questions about water loss but it raised others. For example, because water loss was greatly diminished when leaves were removed from a shoot, such as of laurel, the potometer proved water was lost almost exclusively from leaf surfaces and not from the surfaces of other parts of the shoot. However, it revealed little about the exact site(s) of water losses from leaves, or about the possible role of stomata. Although Erasmus had

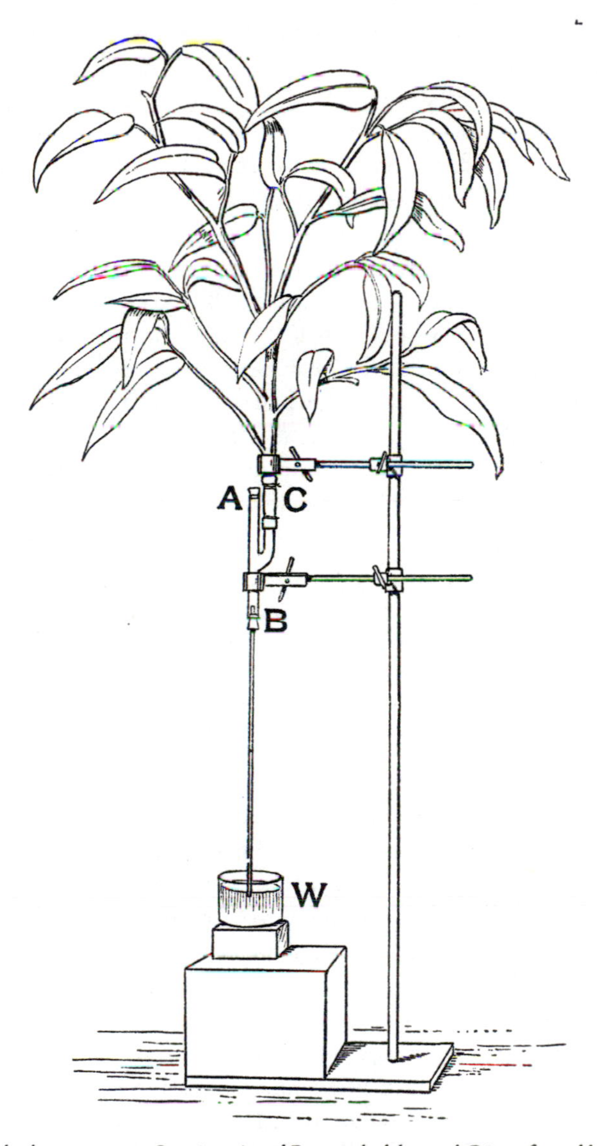

Figure 8.1: A simple glass potometer. Openings A and B are corked, but cork B is perforated by a 20 cm long capillary tube. The cut end of a well illuminated branch of, for example, laurel is connected by rubber tubing to opening C. The apparatus is filled with water before trapped air is released via A. The bottom of the capillary tube dips into a reservoir of water, W. After a period for the plant to equilibrate to the conditions, W is temporarily lowered to allow a small bubble to be drawn into the capillary. The rate of rise of the bubble over a fixed distance can then be timed. The last steps can be repeated at intervals as new bubbles are introduced. Reproduced from F. Darwin and E. H. Acton, *Practical Physiology of Plants* (Cambridge: Cambridge University Press, 1894).

been well aware that leaves have what he concluded were breathing pores in their surface, the detailed nature of these pores had been a mystery to him, and to his contemporaries. Only in the 1820s was it was firmly established that each stomatal pore is surrounded in the epidermis by two specialized 'guard cells'. It had been established by microscopy that such pores were sometimes wide open and sometimes closed, and Sachs had suggested in his *Vorlesungen uber Pflanzenphysiologie* (1882) that stomata acted collectively as 'organs of transpiration' despite their minute individual size, but clearly there was a requirement for a device to measure their opening and closing.

In an early attempt to do this, Stahl in 1894 made use of colour changes in filter paper soaked in 5 per cent cobalt chloride. The paper is blue when it is dry or briefly attached to a leaf surface that contains no stomata, such the upper epidermis of a laurel leaf. It turns pink when wetted or attached to an illuminated leaf surface containing stomata, for example the lower epidermis of a laurel leaf. Stahl's method was qualitative but not quantitative. In *Rustic Sounds*, Francis recalled how, by accident, he found an improved method:

> finding in an old house in Wales a Chinese figure of a man, cut out of a thin shaving of horn, which writhed and twisted when placed on the hand. It was very clearly sensitive to moisture, and it seemed that horn shavings might be used to test the condition of the stomata. The first difficulty was to obtain a supply of this material. Having discovered from the Post Office Directory that there were two horn-pressers in London I proceeded to visit one of them ... He turned out to be of a highly suspicious disposition, but his wife had more discernment, and persuaded him that I was a harmless customer, with no designs on trade secrets, and I finally obtained what I wanted.[28]

A delicate strip of horn, bearing a fine needle, was fixed to a little block of cork, with a graduated scale attached to it, and placed on a leaf. The horn coiled and the needle moved along the scale as humidity increased. Measurements were crude but Francis's horn hygrometer gave the first semi-quantitative measurements of stomatal opening. The hygrometer 'averaged' the opening of all stomata in the area of leaf that it covered but, in retrospect, this was no bad thing because not all stomata in a leaf are open to the same degree at the same time – hence observations of single stomata, as made by microscopy, can be misleading as well as extremely time consuming. It was now possible to distinguish between fully open, partially open, and closed stomata. In 1898, Francis was able to send to the *Philosophical Transactions of the Royal Society* a lengthy paper, 'Observations on Stomata'.[29] He reported stomata close when a leaf is excised from a plant or an attached leaf is placed in dry air. CO_2 caused stomata to close, but they opened in an atmosphere without CO_2. Stomata were more widely open in sunshine than in diffused light; they closed at night or in artificially imposed darkness. He suggested that assimilation (photosynthesis) requires stomata to be open more widely than is necessary for respiration and that, therefore, economy in regard to water is practised at night. And, finally, he noted that opening and closing are periodic phenomena (although less marked than nyctitropism). Rarely in the history of plant physiology can such a simple apparatus as Francis's potometer have revealed so many fundamental truths.

Details were published in 1911 of Francis's even more important invention – the viscous flow porometer – which for the first time enabled stomatal apertures to be quantified, but in order to appreciate why it was so important it is necessary first to examine the contemporary work of Francis's young Cambridge colleague, Frederick Frost Blackman.[30]

*****

Blackman was yet another man who read medicine (at St Bartholomew's Hospital, London) before turning to botany. A brilliant student who entered Cambridge in 1887, he had by 1891 been appointed Francis's Demonstrator – effectively, in charge of his laboratory classes. Francis, who had been promoted to a Readership, communicated the young Blackman's first research papers to the Royal Society, in 1895, and throughout their lives they maintained the closest working relationship.[31]

Blackman, like Francis, had a penchant for designing his own pieces of apparatus to solve specific problems but, in his case, the apparatus was usually complex, often to an extraordinary degree.

> Simplification of technique by complication of apparatus has been the guiding principle and the result is that, although the whole consists of at least eight separate pieces of apparatus, many of them being further in duplicate and all connected together by a plexus of tubes, yet the working is so automatically arranged that the operator, beyond reading the burettes and occasionally working a finger bellows, has nothing to do but turn stopcocks.[32]

Such large laboratory rigs required extensive testing and calibration, something to which Blackman was happy to devote unlimited amounts of time for he was both thorough and cautious by nature. In keeping with his character, he was unusually reluctant to publish – to the frustration and sometimes detriment of his students – but, when he did, his papers always demanded attention.

For those who followed, Blackman had shown there was 'a very remarkable degree of correspondence between the amount of CO_2 liberated [in respiration: or, alternatively, taken up during assimilation] by the two surfaces and the number of stomatic openings per unit area of these same surfaces'.[33] This conclusion contradicted the work of Boussingault, generally accepted at the time, which held that for CO_2 the major pathway into the leaf was through the cuticle covering all its surface, except the stomatal pores. Boussingault had found that leaves of *Nerium oleander* (Oleander) assimilated less when the upper astomatous surface was coated with an 'unguent' (lard) than when the lower stomatous surface had been so coated. Blackman argued conclusively that Boussingault was misled by his use of extremely high concentrations of CO_2 (30 per cent), for while the gas at this concentration penetrated freely through the open stomata its high internal concentration largely inhibited assimilation, whereas in the leaf with stomata almost completely blocked, the relatively low internal concentration still allowed a little assimilation to proceed.

Away from Cambridge, the main centre for physiological research in late nineteenth century Britain was the Jodrell Laboratory at Kew. Since its charitable foundation in 1876, it had attracted a steady stream of visiting workers who could take advantage of its excellent practical facilities and intellectually stimulating atmosphere. It was here that two chemists from the brewing industry, Horace Taberrer Brown, previously known as a pioneer of plant enzymology, and his collaborator F. Escombe, confirmed and extended Blackman's findings. Their apparatus was not unlike Blackman's but in each experiment they were able to enclose much larger areas of leaf. Applying the laws of diffusion and physical models, both to leaves and artificial diaphragms perforated by small apertures to simulate leaves, they concluded that in

> a typical herbaceous leaf ... [the epidermis] illustrates in a striking manner all the physical properties of the multi-perforate diaphragm, which, with its minute apertures set at from 6 to 8 diameters apart, and representing only 1 to 3% of free air, yet allows a perfectly free interchange of gases on its two sides, whilst at the same time affording every protection to the delicate structures underlying it.[34]

Whereas Blackman had confined his interest to the passage of CO_2 through stomata, Brown and Escombe examined also the movement of water vapour – they simply replaced the caustic soda, which they had employed to absorb CO_2, with concentrated sulphuric acid which would absorb the water from air passed over leaves. There was a weakness in their work, however, for they overlooked the importance of air movements in the removal of those 'shells' of water vapour that in still or slow moving air build up around the outside of stomatal pores and whose effect is to slow diffusion away from them but, nevertheless, they had theoretically established the primacy of stomata in the control of transpiration.

Blackman's name will forever be associated with two aspects of photosynthesis, the 'Blackman Reaction' and 'The Law of Limiting Factors'. The first name was slightly inaccurate, arising as it did from work carried out by his student, Miss Gabrielle Matthaei – the 1904 paper (communicated to the Royal Society by Francis Darwin) bears only one name, hers.[35] What was found in this and their joint papers that followed was simple: while at normal ambient temperatures the rate of photosynthesis increases up to a maximum as the intensity of illumination increases, there is little effect of increasing illumination at low temperatures. In the next two decades, it gradually emerged that the overall process of photosynthesis is comprised of two component parts; the first a 'dark' or 'biochemical reaction', limited by temperature like all chemical reactions, and a second or 'light' reaction, independent of temperature. It was the former that the famous biochemist, Otto Warburg, christened the 'Blackman Reaction' (1925). It was appropriate that the evolution of oxygen in the latter reaction was first demonstrated in Cambridge in 1937 by Robert (Robin) Hill, for Hill had as a first-year undergraduate sat through Blackman's lectures.[36]

From the 1904 paper it may seem a short step to Blackman's statement in 1905:

> When a process is conditioned as to its rapidity by a number of separate factors, the rate of the process is limited by the pace of the 'slowest' factor.[37]

Or, as restated by others:

> When the magnitude of a function is limited by one of a set of possible factors, increase of that factor, and of that alone, will be found to bring about an increase of the magnitude of the function.[38]

He had made, however, what was a very large conceptual step forward, whose significance is best appreciated when it is remembered that, up to 1900, leading physiologists like Pfeffer had attempted to interpret the effects of single external factors upon the rate of photosynthesis in terms of so-called 'cardinal points', namely the minimum, optimum and maximum. Their approach was based on the supposition that for any one external factor there existed a minimum level below which the process concerned could not proceed, an optimum at which the rate was greatest, and a maximum level at which the rate was again inhibited. Belief in cardinal points was largely a consequence of investigations that examined only one variable at a time. What Blackman did was to change two or sometimes more factors in a single experiment. As illustrated in Figure 8.2 he could then see that, for example, at light intensity L1 photosynthesis reaches a maximum at a CO_2 concentration of 0.5 per cent, whereas at L3 the maximum is reached at 1.0 per cent CO_2.

The stomatal pathway was clearly the route by which CO_2 entered and exited the leaf, and the role of stomata in regulating transpiration was proven, at least in theory. What was urgently needed now was a means of investigating the factors causing stomata to open and close, or of continuously following stomatal movements over extended periods of time.

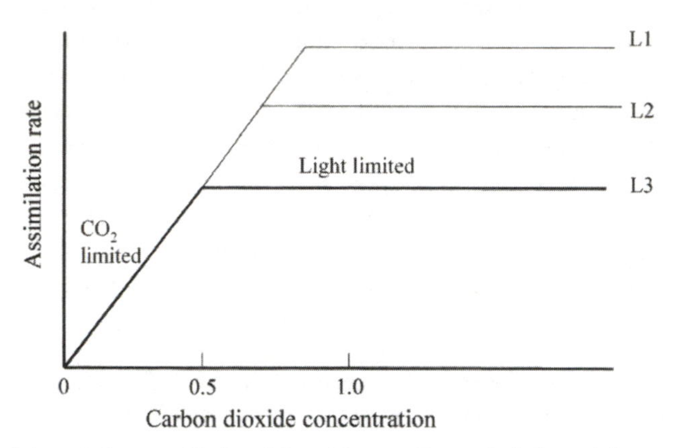

Figure 8.2: A diagram illustrating Blackman's Law of Limiting Factors. At the lowest light intensity, L_3, assimilation increases in direct proportion to rising CO_2 concentration up to 0.5 per cent. Any further increase cannot increase the rate of assimilation, which is now light limited. If, a higher light intensity, L_2, is substituted, CO_2 supply once more becomes the limiting factor: as it is increased, assimilation can again rise proportionately until, at 0.8 per cent, light once more becomes limiting; and so on with L_1. A similar diagram could be obtained by plotting assimilation against light intensity for different levels of CO_2 concentration.

*****

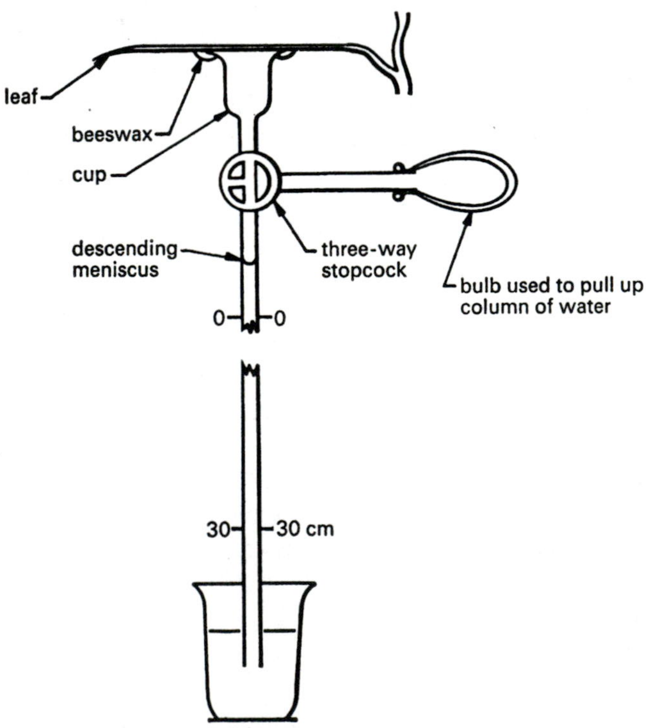

Figure 8.3: The Darwin and Pertz viscous flow porometer. A glass cup was glued to a leaf with beeswax. A column of water in a tube was connected to the cup by a glass T-piece. When suction was applied and the clamp closed, water in the tube descended as it drew air through the leaf. The rate of descent was a function of stomatal resistance to viscous flow. Reproduced by permission of McGraw-Hill, US, from F. Darwin and D. F. M. Pertz, 'On a New Method of Estimating the Aperture of Stomata', *Proceedings of the Royal Society*, B, 84 (1911), pp. 136–54.

The requirement was met by the viscous flow porometer designed by Francis and Dorothea Pertz (Figure 8.3).[39] The principle of the porometer was that it measured the resistance of the leaf to the movement of air that is forced across it, from one surface to the other. Such porometers work best with amphistomatous leaves, i.e. those having stomata in both upper and lower epidermis. Changes in stomatal aperture are measured as changes in resistance and it is assumed that the resistance offered by mesophyll cells in the interior of the leaf – with which, to use an electrical analogy, the stomatal resistance is in series – remains constant. Unless continuously aspirated, CO_2 around the cup attaching the porometer to the leaf can become depleted, which in itself causes stomata to open, but, generally, the instrument was serviceable and a great advance on the horn hygrometer, and on any other device available at the time. Using it, Darwin and Pertz investigated the effects of the humidity of the ambient air upon stomatal aperture and found that opening could be maintained at low humidities so long as the plant was well supplied with water. They confirmed closure at night was followed by opening at daybreak and with their new instrument were able to detect fine details that had escaped earlier investigators. For example, they saw that the stomata of well watered plants opened very wide in the earliest part of the day, only to adjust to a slightly smaller aperture as the day progressed. Movements could for the first time be continuously monitored and their periodicity was starkly revealed.

In the hands of others, the porometer was used to demonstrate that lower than ambient concentrations of CO_2 induced wider than normal opening, while much higher concentrations induced partial closure.

In teaching as in research, Francis made an outstanding contribution to the development of plant physiology, the discipline at the very heart of New Botany. He helped to supply and inspire the generation of graduates that was essential for its continuity and prosperity. There was nothing comparable in the rest of Britain to what was happening in Cambridge. Any impact that the exhausted Vines made on Oxford botany was small, while plant physiology never featured strongly under Francis Oliver at University College, London. In Scotland, 'organography', a form of morphology with some relation to function, was emphasized by Bower in Glasgow and Bayley-Balfour in Edinburgh. Indisputably, Cambridge led the way in plant physiology and in 1917 two of its botanists, F. F. Blackman and his brother-in-law Arthur Tansley, were at the heart of a crusade to put physiology in (what they saw as) its rightful place in the botanical curricula of other British universities, as will be seen in Chapter 11.

When Francis started lecturing in Cambridge, few guides to teaching plant physiology in the laboratory had been published and those managed to be simultaneously unhelpful to the teacher and uninspiring for the student. In the second half of the nineteenth century, the first book to concentrate on practical rather than theoretical considerations of the subject and, therefore, to be of potential relevance to teaching was Sachs's *Handbuch der Experimental Physiologie der Pflanzen* (1865). It was, however, not user-friendly. Experiments were described at great length, the monotony of the 514 pages of text was broken by only 50 illustrations – most of them very small – and there was no index. There was no help either from the two great textbooks that followed in 1868 (Sachs's *Lehrbuch der Botanik*) and 1881 (Pfeffer's *Pflanzenphysiologie*).

The first practical guide to be of any help to teachers of botany was Bower and Vines's, *A Course in Practical Instruction in Botany*. It followed closely the simple but limited courses they had taught so successfully at South Kensington. Consequently the scope was narrow and the focus was very largely on plant anatomy. Part I (1885), which covered the Phanerogamae (seed plants), began with instructions for cutting and staining thin sections of plant material for microscopic examination. There were detailed instructions on how to make up staining reagents, such as Schulze's solution which is used to identify cellulose in cells walls. The names of especially useful plants were given so that, for example, in looking for stomata the teacher was advised to show to students the upper epidermis of a sunflower leaf where 'stomata occur in considerable numbers'. Part II, written by Bower alone because of Vines's ill health and duties in Cambridge, dealt with lower plants, such as mosses and algae, and with the fungi. Neither part dealt specifically with physiology but, accepting the dictum that physiology can be properly understood only if structure is understood, Part I would certainly have given some indirect help to any teacher of plant physiology.

The first truly useful book for physiologists, at least those with a solid grasp of German, was Wilhelm Detmer's *Das kleine Pflanzenphysiologische Praktikum* (1887). In his Preface he wrote

> It is by no means sufficient for the student of plant physiology to attend lectures, or wade through text-books on the subject. He must endeavour to familiarise himself, from practical experience, with methods of research.[40]

In selecting 205 experiments for his readers to adopt in class, Detmer drew heavily on works from German laboratories, such as those of Sachs and Pfeffer, but he included several of his own. Although Detmer carefully illustrated many of the experiments and thoughtfully gave such details as the names and addresses of suppliers from whom seeds, chemicals and apparatus could be obtained, the impact of his book on the students and teachers working in English was probably limited, not just because of difficulties of language but also because by 1898, when Detmer's second edition was translated into English by a S. A. Moor, a Cambridge graduate, a much superior guide, by Darwin and Acton, *Practical Physiology of Plants* (1895), had already appeared. It is interesting, however, to read a tribute to Darwin in Moor's Preface

> The teaching of Plant Physiology in England, which owes so much to the energy and enthusiasm of Francis Darwin, has been seriously retarded by the want of suitable manuals of laboratory practice, and hence no apology is needed for the appearance of a translation of Professor Detmer's well-known and excellent *Praktikum*.[41]

Francis's own Preface sets out the origins of his book

> In 1883 I began a course of instruction in the physiology of plants, of which the chief feature was the demonstration of experiments in the lecture-room. Some years later a different arrangement was made; the students were required to perform experiments for themselves, and at the same time laboratory work in the chemistry of metabolism was organised by Mr Acton. To enable students to carry out their work, written instructions were needed, and the present book is the result of an extension and elaboration of what we prepared for our classes.[42]

So successful was the book that within a year a second edition had been prepared, although Edward Hamilton Acton sadly died before it was published in late 1895. In 1901 Francis revised for a third time the first part of the book, which dealt with General Physiology, and the book was still being reprinted in 1915.

What Francis had done was to reduce to their simplest and most basic form the key experiments in plant physiology conducted in the previous half century. The origins of each experiment are dutifully recorded, so it is possible to see how he adapted from Sachs, Pfeffer and Wiesner, and also from Detmer. His own experiments feature – there are suggestions for experiments demonstrating geotropism and heliotropism and, also, for experiments using the horn hygrometer to measure stomatal movements – as do those of Blackman, whose help with the whole of the physiology section was gratefully acknowledged. Many experiments are associated with clear, simple diagrams; there are instructions for the preparation of reagents and suppliers of materials and equipment are listed. It is noted, for example, that the Cambridge Instrument Company holds stocks of horn suitable for the construction of hygrometers (although Horace's commercial interest is not mentioned!). The book is well indexed and even provides examples of the typical results that might be obtained from certain experiments. In short, it has a thoroughly modern appearance and must have been a boon for teachers and students alike. It is little wonder that the book was so popular.

Its author was popular too with his students for his personal qualities:

> The students who attended his classes felt a deep affection for the teacher who, though not an eloquent or showy lecturer, attracted them by his personality and a modesty which in its sincerity came almost as a shock to the beginner unaccustomed to hear his seniors admit their own ignorance. It was always a pleasure to remain behind after a lecture to ask for help in difficulties; one knew that he would be sympathetic and encouraging; all were made to feel that they were rather fellow students than beginners asking stupid questions.[43]

In his early years in Cambridge, Francis had managed to find time to deal with the demands of teaching and research while initiating, and responding to those many tasks related to his father's work and reputation that are described in the next chapter. His ability to cope with this workload was stretched almost to breaking point, however, when the frustrated Vines left for Oxford in 1888. Francis inherited the problem of having to cope with too many students in too little space. He reorganized the teaching programme – in which Professor Babington no longer took any part – and astutely added a small building extension where practical work could be done.[44] In 1891 it was finally recognized that Babington's decline was irreversible and Francis was appointed Deputy Professor. To add to his problems, Francis did not receive from his colleague, Walter Gardiner, the support to which he was entitled. Gardiner's interest in the department, except for its Museum, had sadly fallen away. He became more and more immersed in Clare College, where he was occupied by his duties as Bursar, and he was of less and less of help to his colleagues in the Botany School. Finally, in the summer of 1895, Babington died. He was eighty-seven years old and he had held the Chair for thirty-four years.

Francis Darwin was the obvious successor. There was every reason why he should be appointed to the premier Chair in British botany. For the last four years he had been *de facto* the head of the botany department, he was a distinguished scientist and popular teacher. Personally, he was likeable, hardworking and trustworthy, and, of course, attached to him was the *cachet* of the Darwin name. But, in the words of Thiselton-Dyer, 'he waived aside his claims to the Chair'.[45] In the autumn of 1895 the Chair was instead offered to, and delightedly accepted by, Francis's old friend, Harry Marshall Ward. Francis had turned away from what would have been seen by others as the pinnacle of a career. Why?

His reasons are not recorded so they must be a matter of speculation, but some are already clear. For one thing, he had within the wider Darwin family taken on the job of arranging, editing and publishing his father's letters, of dealing with new correspondence relating to Charles's scientific works, and of revising his botanical works (see Chapter 9). For another, he must have been acutely aware of the obstacles that were to be overcome if Cambridge botany was to progress. The greatest challenge was how to acquire new lecture rooms, laboratories, and staff offices. In the event, arguing through endless committees and cajoling the decision makers cost Marshall Ward years of effort, frustration and nervous energy before the University was finally persuaded to build a new Botany School.[46]

Family matters too must have added to Francis's unwillingness to go for the Chair in 1895. His mother, Emma, was eighty-seven years old and, although in the close knit Darwin family her care was typically shared between the children, especially those living in Cambridge, her health must have been a serious concern weighing on Francis's mind. At the other end of the scale, his daughter Frances was only nine years old and he wanted to share her childhood in the way that he had not been able to share Bernard's. His enjoyment of married life with Amy had been cruelly cut short by her death, and Bernard had been brought up largely by various nurses and his grandmother before in 1887, at age eleven, being sent away to Summerfield Preparatory School, near Oxford, and later to Eton College. Family life with Ellen and Frances must have been all the more precious to Francis.[47] Ellen may have been an additional concern for, it seems, she was never strong. In 1891–2, for example, she had had to spend part of the winter in St Moritz, 'for her health'.[48]

And then there is the question of hunger. Francis was wealthy and from a distinguished family. He was already fully integrated into that intellectual society which flourished in Cambridge and which was equalled only in Oxford and London. Perhaps he saw few advantages and many disadvantages attached to the Chair. In contrast, his friend Harry Marshall Ward was hungry, and ambitious. He was ambitious not just for himself but, also, to see New Botany established in Britain – he often referred to that objective as 'The Cause'. Marshall Ward had grown up in a lower middle-class home in the newly industrialized city of Nottingham. His father was a poor, albeit respectable, teacher of music. Marshall Ward had risen above his background because he had the talent to win a scholarship to South Kensington where he attracted the support of both a wealthy patron, Louis Lucas, and a powerful guide and counsellor, William Thiselton-Dyer. The Cambridge Chair offered Marshall Ward both a greatly enhanced income and an academically and socially prestigious position.[49]

In refusing to compete with Marshall Ward, Francis was both generous and wise. He saw that what was best for his friend was also best for Cambridge botany. Francis's world was also made happier by his decision for he was freed to concentrate on his own interests.

*****

As the nineteenth century closed, botany, like Francis, was prospering. In recognition of its new status and its increased number of practitioners, botany was in 1895 given its own section, K, within meetings of the BAAS. There had even been a proposal from some animal physiologists that they should form a separate section with the plant physiologists, a proposal that was wrecked by the vigorous opposition of many botanists, led by Bower.[50] Sixty-four years after its foundation, these meetings of the Association were still important for they presented British scientists with the best opportunities, outside meetings of the Royal Society, to exchange news and views, discoveries and gossip.

The first purpose-built botanical institute in Britain, at the University of Glasgow, was opened by Sir Joseph Hooker on 13 June 1901.[51] This was followed shortly by the opening of the new Botany School in Cambridge, on 1 March 1904, by King Edward VII.[52]

At the dawning of the twentieth century, botany was energized, independent and proud of its achievements; it was a fully-fledged member of the family of sciences.

9 FRANCIS DARWIN, FAMILY AND HIS FATHER'S MEMORY

Charles and Emma Darwin had no favourites among their children, excepting perhaps Annie who died when she was only ten years old, but it is difficult to believe that they did not have a special feeling for Francis, their third son, for it was he whose life and spirit was closest to theirs. It was not just that Francis was the biologist among their children. He and his parents shared similar feelings; he shared his father's love of dogs, and he shared his mother's love of music – she played the piano each day at Downe, while he graduated from the penny whistle of boyhood, to the flute and then the bassoon of adult life.[1] And it was Francis who was the first of Charles and Emma's children to marry, and the first to make them grandparents.[2] These first shared joys, followed so soon by Amy's death, forged unique bonds between the parents and their third son. It was fitting then that it was Francis who became his father's biographer.

In the years after Charles's death, Francis presented to the world a sanitized version of his father's life. The first of a series of publications was the *Life and Letters of Charles Darwin* (1887), a book strictly censored by other members of the Darwin family. Edited by Francis, it was based on Charles's autobiography, plus a selection of his letters. Francis also supervised the reprinting and, in some cases, the extensive revision of Charles's publications. The bulk of this work fell into the six years, 1882–8, immediately after his father's death; these were the same years in which Francis was writing and delivering both new lectures and practical teaching in Cambridge, caring for son, Bernard, marrying Ellen and becoming a father for a second time.

The professional lives of academics, particularly of scientists, tend to be calibrated by the dates of their publications, which can be misleading when those lives are viewed in retrospect. There is often a lengthy gestation period between the framing of a question, the conception of an investigation and its culmination in a printed article. Defining pieces of research are sometimes completed long before or after the most productive periods in a life. Given these caveats, there came in Francis's life a period of about ten years from 1888, between his fortieth and fiftieth birthdays, when he contributed little that was new to the literature about his father's life but added significantly to current understanding of the movements of plant shoot and root apices, and, also, established studies of stomatal movements on a firm empirical basis (see Chapter 8). In these precious years, he enjoyed a period of calm achievement, founded on a happy, relatively untroubled, family life. Then there came another storm. Just as Francis was turning his attention back to his father and the task of editing more of his letters, in order to meet the public's insatiable appetite for insights to Charles's life, Ellen's fragile health began to deteriorate. She died in 1903, the same year that *More Letters of Charles Darwin* was ready for publication. Mitral stenosis was diagnosed,[3] suggesting that in her youth she had suffered rheumatic fever, a common condition whose survivors were often left permanently weakened.

*****

Fatherhood and widowhood were the waymarks of Francis's life. It had been the news that he and Amy (his first wife) were expecting a baby that had persuaded the previously reluctant Charles to write an autobiography. For years he had been refusing the pleas of various publishers for such a book but now, he said, he would put something on paper to amuse his grandchildren. He was adamant that what he wrote was not for publication and, therefore, he felt free to commit to paper his thoughts on religion and to record frankly his memories of many distinguished and powerful people. The autobiography had been read by Emma, so its contents were not secret to her, or to other members of the family.[4] Like Charles, however, they expected that the biography would remain private within the family, hidden from the world. Like Charles, they too, naively underestimated public interest following his death. Stung by the news that others would publish their own versions of 'A Life of Darwin' if they did not, the family agreed that Francis should undertake the job of producing an official version, sanctioned by them. It would contain his autobiography, plus selected letters. Charles's publisher, John Murray, was pleased to sell *Life and Letters*, especially as the Darwin family was ready to bear the cost of its printing and binding.[5]

Under pressure from Emma and other members of the family, especially Henrietta, Francis had edited out about 17 per cent of his father's writings. (The exact nature and extent of the excluded material is known because in 1958 his granddaughter, Nora Barlow, published an unexpurgated version, clearly identifying for her readers the previously unseen material). The portions that had been excluded were thought by the family to reflect negatively on either themselves or the reputation of others. Thus, statements that could be construed as critical of other people were omitted, as were sections about Charles's religious beliefs (the latter were a particular concern of the orthodox, god-fearing, Emma). For example, in one of the omitted passages, Charles had written,

> But I had gradually come by this time [viz Oct 1836 to Jan 1839] to see that the Old Testament [from its manifestly false history of the world, with the Tower of Babel, the rainbow as a sign, etc., etc., and from its attributing to God the feelings of a revengeful tyrant] was no more to be trusted than the sacred books of the Hindoos, [or the belief of any barbarian].
>
> ... disbelief crept over me at a very slow rate, but was at last complete. The rate was so slow that I felt no distress, and have not ever since doubted for a single second that my conclusion was correct. [I can indeed hardly see how anyone ought to wish Christianity to be true; for if so the plain language of the text seems to show that the men who do not believe, and this would include my Father, Brother, and almost all my friends, will be everlastingly punished. And this is a damnable doctrine.][6]

Life and Letters ran to three volumes. It was such a success that Emma told a friend, 'Frank says he has lost all modesty ... and I hope it is partly true. His nature is to doubt and disparage everything he does'.[7] Exactly as the family had hoped, *Life and Letters*, being the first published account of Charles's life, created for the public a solid, positive image that later writers would find difficult to challenge. It stood the test of time. Almost forty years later, in 1925, Sir Charles Sheringham was able to reflect in his Presidential Address to the Royal Society, the book was 'by common consent one of the most admirable and delightful accounts ever written of a great scientific life, the modesty and simplicity of the presentation contributing to its charm'.[8]

While he was assembling *Life and Letters*, as well as dealing with day-to-day correspondence, Francis was also busy revising two of Charles's botanical books. The first, *The Different Forms of Flowers on Plants of the Same Species* (see Table 6:1), was republished in 1884, with a Preface by Francis evaluating progress in the subject since the first edition in 1877. A second edition of *Insectivorous Plants* was completed in 1888. Francis had incorporated in his revision Charles's notes made in the margins of the original manuscript and, in order to update the book, he had added many footnotes referring to recent publications – in some instances citing his own. The notes are collectively so extensive that they amount to a commentary on Charles's original text.

More Letters of Charles Darwin, comprising two volumes, followed in 1903. In producing them, Francis had had the assistance of a co-editor, Albert Seward, his colleague in the Cambridge Botany School. The family was happy that an outsider should be involved with these later volumes not only because of Seward's personal probity but because, generally, they were feeling more relaxed about the need they had perceived to protect Charles's reputation. Moreover, by 1903, Emma and a number of Charles's contemporaries had died, so diminishing the risk of irritating any personal sensitivities. Seward's help was needed also because it was clear that, although Ellen's powers of encouragement and criticism greatly helped Francis with his writing,[9] her poor health was occupying more and more of his time and energy.

*****

It was Ellen who, earlier, had proved the key to Francis's period of calm achievement. She provided the stable family life and leading position in Cambridge society that enabled him to devote himself to his science in a way that had not until then been possible. Ellen was not the only graduate of Newnham College who helped shape his life, as will be seen.

The college had been the brainchild of her cousin, Henry Sidgwick, a philosophy don of Trinity College (see Chapter 8). He had in 1871 established in Cambridge a house, or residence, for women who wanted to attend the small but expanding range of lectures available to them. Increasingly, the women of Newnham – similarly of Girton College, which was founded two years before Newnham – were being admitted to lectures given in college halls and lecture rooms around the city. As early as 1871, for example, St John's College had permitted one of its lecturers, Mr Main, to give instruction to women students in chemistry, albeit usually at an early hour, such as 08.30, before lectures to undergraduates began. In response to energetic lobbying by Sidgwick and like-minded dons, twenty-two out of thirty-four professors in the University had by 1873 given their permission for women to attend lectures (the physiologist, Michael Foster, also allowed them to participate in laboratory instruction). By 1875, the College had its own buildings. Young male graduates, such as Harry Marshall Ward, sometimes taught at Newnham on a temporary basis while they waited for permanent employment,[10] but by 1879 the college had established its own list of female staff who, in addition to their tutorial role, had to teach those subjects for which the women were excluded from University lectures. By this point the college had its own buildings that were accommodating thirty students, while a further twenty-five were in lodgings.

This was the world into which the nineteen-year-old Ellen Crofts was admitted in 1884.[11] It was a world of excitement, challenge and innovation, but it was also a male dominated world in which every aspect of the behaviour of the young women of Newnham and Girton was closely scrutinized by supporters and critics, men and women, alike. Ellen was described within the Darwin family as a 'blue stocking', intelligent but also intensely reserved and 'difficult to know'.[12] In fairness to Ellen, her position within the family was always going to be difficult. When she married Francis, her stepson, Bernard, was seven years old. Her relationship with him, like that of all stepmothers with their stepsons, was full of potential problems but it seems that, at least in her husband's eyes, Ellen succeeded in being loving and kind. In letters sent to Bernard's grandmother, Nain Ruck, Francis wrote warmly about the relationship between Ellen and Bernard. He told Nain:

10 December 1883

> His [Bernard's] favourite greeting to Ellen is to bump his forehead gently against hers ... He is always asking how she is.[13]

For his part, Bernard remembered

> Ellen was always kind as could be in reading to me and playing with me, but there was some feeling of reserve; perhaps she tried too hard to be a good stepmother and never to outstep those limits.[14]

Bernard was ten when his half-sister, Frances, was born in 1886. One year later he was sent away to boarding school. He might have felt rejected by his family but in his own account of his childhood there is no hint of this, or of any feeling of resentment towards either his stepmother or his beloved father. Towards his little sister, there were only feelings of love and memories of the happy hours they enjoyed together during school holidays. As they grew older, they read together and made up sonnets, a game in which Frances already displayed the talents that would make her an acclaimed poet in adult life.

Frances spent much of her childhood playing with, and exploring her expanding horizons alongside her two cousins, Gwen (George's daughter) and Nora (Horace's daughter). The three girls were of similar age, being born within a year of each other. Gwen looked upon her Aunt Ellen as sophisticated and rather shocking – in an admirable way. Ellen's house, Wychfield, was a jumble of lovely objects, Persian rugs and brightly coloured cushions mixed up with golf clubs and tennis rackets. Everywhere were bowls of water for Uncle Frank's beloved dogs, and Frances's donkeys roamed freely through the large garden. In summer, Aunt Ellen might be found sitting on the veranda playing word games with her friends and, most shocking of all for Gwen, smoking cigarettes. Among the smokers might be found Jane Harrison; she and Ellen had been best friends ever since they were admitted together to Newnham in the cohort of 1884. A brilliant classical archaeologist, Jane's life was blighted by unhappy love affairs with both men and women. On graduation she worked at the British Museum and lectured – dazzlingly – in London before accepting a research fellowship at Newnham in 1898.[15] Pushy, self-dramatizing and self-promoting, Jane Harrison more or less invented the role of professional female academic as a researcher as well as a teacher. She successfully challenged the way classicists thought about the culture of ancient Greece and whether those in the close-knit academic community of Cambridge loved or hated her – and many fell into the latter group – she could not be ignored.

Ellen continued to be involved in the governance of Newnham after she married and gave up her lectureship. It was through Ellen and such friends as Jane, that Francis was drawn into a social circle of artists and classicists. His naturally wide interests and tastes could hardly fail to be stimulated, but he remained at heart a scientist. In her biography, Jane wrote:

> One scientific friend, Francis Darwin, had a lasting influence on me. Classics he regarded with a suspicious eye, but he was kind to me. One day he found me writing an article on *Mystica Vannus Iacchi*. 'I must get if off tonight', I said industriously. 'What is Vannus?' he asked. 'Oh a fan', I said, 'a mystical object used in ceremonies of initiation'. 'Yes, but Virgil says it is an agricultural implement. Have you seen one?' 'No', I confessed. 'And you are writing about a thing you have never seen?, groaned my friend; 'oh you classical people'.
>
> [Shortly afterwards] Francis Darwin unearthed a mystic 'fan' in remote provincial France. He sent it to Cambridge.[16]

Jane was extremely fond of Francis, regarding him as 'an old bear', and he was equally fond of her. During term times she was a regular guest at Sunday lunches at Wychfield and she sometimes joined the family on their holidays so that, in adulthood, Frances could 'hardly remember a time when "Aunt" Jane was not part of my existence'.[17]

One direction in which Francis was pulled by his wife, though not unwillingly, was into the fight for women to be recognized as full members of the University.[18] The struggle for sexual equality proved fruitless during their lifetimes but it did motivate a number of young female botanists, as will be seen.

Although in 1881 women were allowed for the first time to take the Tripos examinations, the arrangements to sit had to be negotiated individually with each examiner. Most importantly, women could still not receive degrees, or even use the University Library. Lists of their results would be published separately, although alongside those of the men, and each would obtain a certificate stating how she had performed. Six years later, in 1887, there was the first concerted attempt to persuade the University to allow 'titular' degrees to be conferred upon women. The suggestion was modest; women would be given degrees in name alone, and it was acknowledged they would still not play a full part in the life of the University. The proposals were defeated. They were defeated, again, and much more publicly, in 1897. Objectors plastered the town with posters declaring 'Down with Women's Degrees' and 'Frustrate the Feminine Fanatics', and they hired special trains from London bringing to vote at the Senate House large numbers of sympathetic MAs – men who had voting rights by virtue of having a Masters degree. The motion was crushed and a jubilant and boisterous crowd of undergraduates vandalized Market Square before marching on Newnham. Fortunately, for everyone's sake, the Principal, Eleanor Sidgwick had closed and barred the gates. Yet again, in 1921, the status of women was debated. A motion to allow them full degrees was rejected, although they were henceforward to be granted titular degrees. This time the male undergraduates celebrated victory over women by using a handcart to batter down the lower half of the Newnham gates.[19]

Full equality had to wait until 27 April 1948 when, by Order in Council of King George VI, women were finally admitted to full membership of the University of Cambridge. The struggle for equality had been bitter and protracted but, against this background, the nineteenth century had seen remarkable changes in the education, self-esteem and aspirations of women. Their struggle was interwoven with the struggle of botany to escape from eighteenth-century thinking. One of the problems of botany in the late eighteenth century had been that, although it was seen as a subject fit and proper for women to pursue, it was thought that their activities should be restricted to collecting and drying, naming and drawing plants (see Chapter 3).

The attitudes of women, let alone men, changed little until the middle of the nineteenth century, at the earliest. The status quo was perpetuated because few young women had access to formal secondary schooling; at best they were tutored in their homes. Newnham recognized that many of its first students were thus unable to cope with degree work and tailored its teaching accordingly, so that some students took an additional foundation or preparatory year to help them cope with the level of University courses (Girton was less flexible and, as a result, attracted fewer students). The establishment of women's colleges in Cambridge, Oxford and London, initiated a virtuous circle. As more women were educated to graduate level, so many became schoolteachers and role models for the next generation. There was in the last quarter of the nineteenth century an upsurge in the number of girls' secondary schools that were founded. Botany was a popular subject for young women both in these schools and in universities, perhaps reflecting the vestiges of eighteenth century attitudes. However, the girls learned a very different subject from that familiar to Rousseau or Lindley. They were increasingly exposed to the spirit of New Botany and began at last to make an original contribution to the science of plants.

Several of the first graduates of the new women's colleges found their way into research posts in the Botany School, two of them collaborating with Francis. One was Anna Bateson, who was involved in his studies of gravitropism.[20] In a pedestrian paper, they confirmed Sachs's teaching that organs placed horizontally, rather than obliquely, show the greatest geotropic response. And, in a similarly uninspired paper, they explored the effects of a range of chemicals, some of them, such as ether and camphor, clearly toxic, on turgescence and the growth of pith from Jerusalem Artichoke (*Helianthus tuberosus*) and Sunflower (*H. annuus*).[21] Anna subsequently published on the same subject on her own. The daughter of the Master of St John's College, Anna had entered Newnham in 1882, where her mother was a member of the College's Council from 1880 to 1885, and there she read botany. She was, like her brother William the pioneer of Mendelian genetics, both headstrong and independent. Unable to settle in academic life, and lacking the originality of her brother, she eventually quit Cambridge in 1890, starting her own market-garden near Bournemouth, an occupation for which the big, strong, beer-drinking, pipe-smoking Anna was much better suited.[22]

A contemporary of Anna's at Newnham, and a more talented scientist, Dorothea Pertz was Francis's other main female collaborator. German-born, she was ten years younger than Francis although, possibly, he had known her from childhood since she was the niece of Charles and Mary Lyell, close friends of Charles and Emma Darwin. She certainly retained a youthful memory of visiting the Darwins at Down – where, to her lasting delight, she had been allowed to caress the nose of the horse upon which sat the great Charles Darwin. Dorothea's work with Francis spanned ten years and several subjects. It culminated in their groundbreaking paper on stomata, published in 1911 (see Chapter 8). Although she also worked with William Bateson, studying variation in the flowers of *Veronica*, Dorothea felt she lacked the mathematical and chemical skills needed to pursue an independent research career in plant physiology after Francis retired.[23] She did, however, make other contributions to botany, at the request of F. F. Blackman using her native ability in German to index several journals published in that language. She also used her outstanding artistic skills to illustrate various botanical publications, including a series of papers on floral morphology written by her friend Edith Saunders. Francis's niece, Nora, was not a Newnham woman but she too was a botanist, studying plant physiology with Blackman and genetics with Bateson. Before her marriage to Alan Barlow, she had taken up one of the subjects that had occupied her grandfather, heterostyly; like Charles she tackled the mysteries of *Primula* but she then moved on to *Oxalis* and *Lythrum*, publishing two papers in the *Journal of Genetics*. Nora represents the fifth generation of Darwins for whom botany played an important part in their life.[24]

While Francis collaborated with Anna and Dorothea, and, at about the same time, Harry Marshall Ward was collaborating with Dorothea Marryat and Edith Dale (a Girton graduate), it was Edith Saunders who was the first of Cambridge's female botanists to make a major, independent impact upon her subject. Edith entered Newnham in 1884, two years after Anna and Dorothea, and was later made a Fellow of the College, organizing in a converted chapel some of the first practical work that women students could attend. Making use of a large plot rented in the University's Botanic Gardens, she took an experimental approach to investigate evolutionary processes – which would almost certainly have found favour with Charles Darwin. Indeed, her earliest studies of intercrossing in *Biscutella laevigata* (a small Brassicaceous plant from the alps) might have been designed by, and were certainly worthy of, the great man. Using seed that Bateson had brought back from a trip to Italy and studying the inheritance of contrasting characters – smoothness and hairiness of leaves – she was able to distinguish what would soon, when Mendel's work was rediscovered in 1900, be called dominant and recessive characters. Her findings were published in 1898 in the *Proceedings of the Royal Society*. In 1902, Bateson and Saunders reviewed for the Evolution Committee of Royal Society, in the first of a series of five Reports, the extent to which available data on inheritance agreed with Mendel's principles. Saunders's cross-breeding experiments with *Lychnis*, *Atropa*, and *Datura* (funded by a small grant that Bateson had obtained from the Royal Society) were pivotal since they were clearly consistent with those principles and important pieces of evidence supporting them. In their landmark report, Bateson and Saunders used several terms that have since been integrated into the basic language of genetics, terms such as allelomorph (now allele, meaning one of the two or more alternative forms of a gene), homozygous and heterozygous (having, respectively, identical or different alleles at the locus on homologous chromosomes), and P (parental), F1 and F2 (filial) generations.

After 1902, she and Bateson were joined by other female workers, including Florence Durham (Bateson's sister-in-law) and Muriel Wheldale. Indeed, of the thirteen researchers linked with Bateson's research, no less than seven were women, all them associated in one way or another with Newnham or Girton.[25] Not only were women taking their place at the cutting edge of botany, they were consciously setting out to prove the scientific credentials of their sex. Women found the best chance of employment in the younger, 'nascent' disciplines, such as plant physiology and genetics, because men did not, as yet, see in those subjects safe, well-funded career paths.[26] As an avid supporter of women's rights, and secretary of the committee of dons who supported the 1897 campaign for Degrees For Women, Bateson was only too happy to use the talents of the Newnham women.

Two female graduates of the Botany School, although closely associated in the early years of their careers with Bateson's investigations of pigments in flowers, went on to find fame in their own right. Florence Wheldale combined genetics with chemical analysis and, partly on the strength of her landmark book, *The Anthocyanin Pigments of Plants* (1916), was awarded a lectureship in biochemistry, becoming one of the first women to be appointed to a science department in the university.[27] Edith Saunders remained a classical geneticist but she too developed into an independent and original scientist of the highest calibre, as was recognized by the Linnean Society which made her one of its first female Fellows in 1905 and, towards the end of her career, by the Genetical Society which made her its President from 1936 to 1938.

*****

To return to Ellen, and this time her death, it was at about the turn of the century that her health declined. She needed a major operation in 1901 and in September 1903 she died. She was only forty-eight years old; Francis was fifty-five. He was devastated. For the second time in his life his wife had been taken from him. For the second time, he was left alone with their child. In some ways his parental role was more difficult than when Amy died. Whereas Bernard had been only a few days old when his mother died and had never known her love, Frances was seventeen when Ellen died and she had been very, very close to her mother. Whereas, Emma Darwin, Nain Ruck, and a host of aunts and uncles had been there to try to make up to Bernard for the loss of his mother's love and attention, there were this second time around no grandmothers to support Frances, and the Darwin aunts and uncles were needed by their own families. What support Frances received came from her cousins, Gwen and Nora, and from old family friends of her parents, especially Aunt Jane and Will[iam] and Alice Rothenstein. Inevitably though, the major burden fell upon her father.

Heartbroken, Francis and Frances did not attend Ellen's funeral. Instead they walked into the country, found a quiet wood and, in keeping with Ellen's spirit, read Jane Austen's *Sense and Sensibility* to each other. As soon as was possible, Wychfield was sold and father and daughter fled to London to escape their empty home. 'I started vehemently working at an art school [Westminster]. Then I broke down', wrote Frances.[28] The breakdown, in 1904, was to be the first of many. Her mother had been prone to bouts of depression – as too was her father – but Frances's periods of illness were lengthy and severe, occurring throughout her adult life. Those following Ellen's death, ultimately, caused her father to sacrifice his career. Torn between the needs of his family and his wish to continue supporting his ailing friend and professor, Marshall Ward, he was soon forced to recognize that he could no longer fulfil reliably his teaching and administrative duties in the Botany School. On 28 August 1904, Francis resigned the Readership he had held for sixteen years.[29]

Frances was sent to recover at 'Jodulhutte', a small convalescent home high in the forests near Bex, in Switzerland. But her recovery was slow and punctuated by relapses.[30] She remained in Switzerland for most of the next three years, visited regularly by her father and a series of cousins who kept watch over her. The love she received, together with invigorating mountain walks, finally gave Frances a little more energy. With Gwen's encouragement, she began to paint again and through poetry she began to release the powerful emotions trapped inside her. But still she was fundamentally unwell and it was decided that she should return to England for 'an old fashioned rest-cure'. Frances tells how at last she recovered

> Here, the miracle happened ... this resurrection was, I think, the most unexpected, extraordinary period of my life. My father had now come back to Cambridge and I came home at Christmas perfectly well ... Naturally I turned to [Aunt Jane] at once. She said she would teach me Greek. For two terms she taught me unofficially at odd times, elementary Greek ... She promised to let me draw her: I was still as determined as ever to be a painter. Every day was full now and seemed like a fresh melody, it was so happy.[31]

Aunt Jane did Frances one other service – to her own cost. She introduced her to a young classics don, Francis Cornford, with whom she, Jane, had enjoyed the closest of friendships. (Characteristically, Jane was probably deeply in love with the younger man who never regarded the relationship as other than Platonic).[32] Frances and Francis Cornford fell in love and were married in 1909, shortly after the festivities to mark the centenary of her grandfather's birth. The bride's father employed his friend, the distinguished architect Arthur Clough, to design a house for the young couple and had it built, at his expense, on Madingly Road, Cambridge. Eric Gill carved in the corbel in the hall, 'Francis Darwin built this house for Francis and Frances Cornford 1909'.[33] Jane Harrison took time away from Cambridge to get over her latest lost love before finding peace and happiness with her final love, Hope Mirlees.

Among those who had supported and inspired Frances were the Rothensteins. Will was a portrait painter – later, a war artist and Principal of the Royal College of Art but, always, a socialite. His connection with the Francis Darwins stemmed from mutual friendships with Ellen's brother, Ernest – a popular painter of historic battle scenes – and was reinforced through Will's commission to paint a portrait of Francis to mark his retirement. On staying with the Darwins when Frances was sixteen, Will and Alice had guided her around the Fitzwilliam Museum in Cambridge and, also, the wonderful art collections held in several of the colleges. With words of encouragement they had given her chalk, crayons and a sketch pad. In the tragic year of 1904, the couple had taken her to art galleries in London, where they lived, and given her the confidence to paint with Will in his studio.[34]

Will wrote 'she was a sort of pupil of mine ... She both drew and wrote poems with a simple sincerity, and was wise beyond her years'.[35] He was the first person to suggest that she had an exceptional talent for poetry and it was he who later persuaded her father that Frances's poems should be published 'at all costs'.[36] Thirty pounds, in turn, persuaded a bookseller in Hampstead to publish this, her first collection of poems. As her fame grew, several of these early poems were reprinted in subsequent volumes. Almost as soon as Frances had become engaged to Francis Cornford, her father wrote to Will breaking the glad news and inviting him to join the trio at Lyme Regis. This was not the only holiday they were to spend together, for Francis's friendship with Will lasted through the rest of his life. Francis helped Will get commissions for portraits from such scientific worthies as Alfred Russell Wallace and Joseph Hooker, and introduced him to many of his Cambridge colleagues, including William Bateson with whom there was immediate empathy. The two Williams certainly shared an interest in art, for Bateson was a knowledgeable collector of fine art with a special interest in Chinese and Japanese works, but, over and above this, the two men and their wives simply enjoyed each others' company. Thus began a triangle of friendship between the Darwins, Rothensteins and Batesons.

That Francis Darwin and William Bateson were friends is remarkable, for Bateson was the leader of what was regarded at the time as an anti-Darwinian movement. He was, moreover, an outspoken man, seen by his opponents as disputatious and aggressive. It says much for the character of each man that he was able to put aside any scientific differences in the greater interests of their friendship.

A zoologist by training, Bateson had travelled and studied widely before settling in Cambridge and publishing in 1894 his *Materials for the Study of Variation Treated with Special Regard to Discontinuity in the Origin of Species*. In this work he pointed to the discontinuities between species rather than the continuities emphasized by Charles Darwin. For Charles's supporters, 'blending' had become an increasing problem, for they found it increasingly difficult to defend the common observation that offspring tend to be a blend of their parents' characters. It was difficult for them to explain how variation –the raw material for natural selection – quickly disappears as differences are smoothed out over successive generations. Bateson argued that the evolutionary process was made up of a series of jumps, or 'saltations', rather than small increments. Interestingly, while Charles had tended to associate discontinuous variation with the production of monstrosities and had argued that complex new adaptations could not possibly be acquired by a sudden single jump, Huxley, normally his staunchest supporter, had been moved to write to *The Times* (April 1860)

> Mr Darwin's position might, we think, have been even stronger than it is, if he had not embarrassed himself with the aphorism '*Natura non facit saltum*' ... We believe ... that nature does make jumps now and then. And a recognition of the fact is of no small importance.[37]

When Hugo de Vries – who by his studies of *Oenothera lamarckiana* (Evening Primrose) convinced himself that such jumps were the result of mutations – sent to Bateson in April 1900 a copy of an overlooked article published in 1866 by Gregor Mendel in the obscure *Proceedings of the Natural History Society of Brünn*, Bateson saw the evidence he needed to prove his saltatory vision. Mendel's peas were either smooth or wrinkled. Inheritance was by irreducible particles; it was non-blending.

Bateson's first practical contribution was to show that Mendel's Laws applied to animals as well as plants. He made Mendel's works more accessible by translating them into English. He is most often remembered, however, for a series of attacks, launched most publicly at the BAAS meeting of 1904, against a group known as the 'biometricians' who were led jointly by his former mentor and Cambridge colleague (WF) Raphael Weldon and, from University College London, by Karl Pearson.

Bateson's books, *Mendel's Principles of Heredity: A Defence* (1902) and *Mendel's Principles of Heredity* (1909) were enormously influential. He coined the word 'genetics' and, in collaboration with Edith Saunders, many others in the modern genetical lexicon, as already noted. As also mentioned before, he attracted around him a group of extremely talented young men and women. Studying the genetics of *Lathyrus odoratus* (Sweet Pea), together with Saunders and their talented colleague Reginald Punnett, he discovered in 1905 that some genes do not segregate independently during meiosis (the type of cell division that occurs in the sexual process); instead they are linked in some way. This was of enormous importance because the lack of total independence suggested that the greater the frequency of linked inheritance, the closer two genes are to each other physically. Acceptance of the principle opened the way to the quantification of linkage and construction of genetic maps but, in perhaps the greatest error of his career, Bateson denied almost until his death what Thomas Hunt Morgan had discovered, namely that genes are physically located within linear structures called chromosomes. While Saunders was quickly convinced by the strength of Morgan's evidence, Bateson was won over only after he had seen the evidence at first hand when he visited Morgan's laboratories in Toronto in1921.[38]

For many years Bateson struggled along in Cambridge as deputy to the professor of zoology, poorly paid and constantly battling for research funds. In 1906 he was elevated to the position of Reader and in 1909 a Chair of Genetics was created especially for him, thanks to the generosity of an anonymous donor[39] – who it was widely believed was Arthur J. Balfour. However, the funding was for a fixed term of only five years. It proved not enough to keep him in Cambridge and he left in 1910 to become the first Director of the John Innes Horticultural Institute in Merton, Surrey. He was disillusioned by years of being underpaid and struggling to resource his laboratory. He had occasionally received support from others, as when Francis Darwin tracked down the £200 Bateson needed to build an experimental glasshouse at the University Botanic Garden, but too often he had had to make do with growing plants in his own garden (Edith Saunders often used the gardens of Newnham College) and his battles had to be fought alone.[40]

Passionate argument between Bateson and his 'Mendelians' and the biometricians raged on through the first two decades of the twentieth century.[41]

Personal animosity was never far below the surface and, as so often happens in such disputes, each party wilfully misrepresented the views of the other and was selective in weighing the evidence of the other. Thus, for example, the biometricians chose to ignore a passage in Mendel's writings where he suggested that some of his particles [of inheritance] could probably have even very small, and interacting effects, which, of course, was consistent with the idea of continuous variation. The differences between the positions of the two camps did not start to erode until 1918 when, in a seminal paper, Ronald (RA) Fisher demonstrated that continuous characters were each governed by several independent mendelian factors, every factor or gene having only a small phenotypic effect (or physical outcome). Such characters are said to be 'polygenic'. Interestingly, Fisher found support for his views from another of Charles's sons, Leonard, who, in their long exchange of letters, attacked Bateson in a way that his brother, Francis, would never have done. Leonard's assertion that 'nothing but harm can come from following Bateson in regard to evolutionary theory'[42] was echoed by Fisher, whose stark judgement was that Bateson had greatly retarded the progress of genetics in Britain. All this about a man who had been awarded the *Darwin* Medal of the Royal Society in 1904!

As more and more was revealed about the behaviour of chromosomes and individual genes during cell division, it became evident that mutations – changes to the properties of individual genes – were much less common than de Vries had supposed. Modern re-analysis shows that only 2 out of the 2000 mutations he claimed would today be recognized, the others were, simply, recombinant genotypes, i.e. were due to the physical rearrangement of *unaltered* genes. The enormous store of variability latent in sexual organisms has become clearer – so reducing the necessity of a role for mutations – and evidence has accumulated that physiological traits, such as drought resistance, as well as structural characters, such as height, are under polygenic control and are continuously variable. The many genes of small effect, regulating a single character, may be aggregated into a few separate 'Quantitative trait loci', often distributed between several chromosomes.[43]

Despite advances in molecular genetics, much scope remains today for argument and debate about many of the topics that occupied Charles's thoughts, such as the role of geographic isolation in speciation. For saltationists, there was no need for isolating mechanisms to restrict gene exchange; new species arose suddenly, and old and new species could exist sympatrically (within the same geographic region). With the triumph of a view of evolution that is largely gradualist, the weight of opinion has turned back in favour of the need for geographic separation in speciation, i.e. in favour of allopatric speciation, but the matter is far from settled.

*****

There were few aspects of Charles's life about which either his contemporaries, at least those outside his family, or historians could be critical. One exception, however, was his behaviour in the 'Darwin/Butler Controversy'. When he did make brief reference to the dispute, Francis was, once again, at pains to minimize any damage to his father's reputation but, in this endeavour, Francis probably had another motive, which was to salve his own conscience.[44]

Friendship between the Darwin and Butler families stretched back for generations. It had begun with Charles's father, Robert, the general practitioner and leading citizen in the small market town of Shrewsbury, and Dr Samuel Butler, the headmaster of Shrewsbury School and an equally prominent figure in the town. Among Charles's classmates at the School was the headmaster's son, Thomas, and the two were reunited as undergraduates in Cambridge. They spent the summer of 1828 together on a reading party at Barmouth, North Wales, and Butler said of Charles, 'He inoculated me with a taste for botany which has stuck with me all my life'.[45] This Butler was the father of the Samuel Butler who proved to be such an irritating thorn in Charles's flesh during the last years of his life.

Long before Samuel Butler wrote his classic novel, *The Way of All Flesh*, he began his working life as a sheep farmer in New Zealand and it was there that, with considerable enthusiasm, he read the *The Origin of Species*. However, by the time he returned to Britain in 1864 his enthusiasm was waning. When some critics saw in his novel *Erewhon*, published in 1872, an attack on Charles's theory of evolution he was quick to write to Charles saying, 'I am extremely sorry that some of the critics should have thought I was laughing at your theory, a thing I never meant to do'.[46] Soon after this, he twice visited Down House, meeting most of the family. Relations were obviously cordial at this stage and he and Francis seem to have become good friends. In 1877–8, in particular, they met quite frequently, discussing their shared interest in the role of the mind in evolution. Butler was increasingly worried that Charles's theory of evolution gave too much prominence to accident and not enough to design. Butler favoured the theory of 'use-inheritance', which proposed that the most useful, or used, features are the most likely to be inherited. Finally, he became convinced that the mind is the controller of evolutionary direction, essentially aligning himself with a number of eighteenth century thinkers, including Erasmus Darwin. It was Erasmus who proved to be the spark that lit the tinder of controversy.

To celebrate Charles's seventieth birthday in February 1879, the German language journal *Kosmos* published an article by Dr E Krause on Erasmus's contribution to evolutionary theory. In May of the same year, Butler published *Evolution Old and New, or the Theories of Buffon, Dr Erasmus Darwin and Lamark Compared with that of Mr C. Darwin*. Events moved swiftly so that, with Charles's help, Krause was soon enlarging his essay for translation and publication, which was to be included as part two of Charles's own *Life of Erasmus*

Darwin. Charles sent a copy of Butler's work to Krause who, in his revising his part, made several disparaging remarks about Butler's ideas (without specifically referring to Butler's book). Unfortunately, in the Preface to his *Life of Erasmus*, although Charles noted that *Evolution Old and New* had appeared subsequent to Krause's article, he failed to record that Krause's contribution had been revised after the publication of Butler's work. Butler compared the original German and revised English versions of Krause's essay and saw a covert attack upon himself. He concluded that Charles had conspired to present Krause's work as independent and pre-dating his own *Evolution Old and New*. Butler was wounded and angered. Writing to Charles on 2 January 1880, he demanded an explanation. As ever, Charles dealt promptly with his correspondence; the next day he replied, offering both explanation and regret to Butler:

> 3 January 1880
>
> My dear Sir,
> Dr Krause, soon after the appearance of his article in *Kosmos*, told me that he intended to publish it separately and to alter it considerably, and the altered MS. was sent to Mr Dallas for translation. This is so common a practice that it never occurred to me to state that the article had been modified; but now I much regret that I did not do so.[47]

He was mistaken if he thought the matter would rest there, for Butler continued to pursue it. He made the dispute public through letters he sent to the *Athenaeum*. Charles too was now wounded. Always vulnerable to personal attacks, he was hurt, confused and uncertain how to respond. He drafted letters of explanation and refutation but on the advice of Huxley, amongst others, he did not send them. Butler did not give up and the affair dragged on past the deaths of both men (Charles's in 1882, Butler's in 1902).

When Francis edited his father's *Autobiography*, to be included in the first volume of *Life and Letters* (1887), he omitted all reference by his father to the quarrel with Butler. In the third volume, however, Francis does refer to the affair, saying there was an attack upon his father which amounted to a charge of falsehood, but

> After consulting friends, he [Charles] came to the determination to leave the charge unanswered, as being unworthy of notice ... The affair gave my father much pain, but the warm sympathy of those whose opinion he respected soon helped him to let it pass into a well-merited oblivion.[48]

And there the matter rested until 1911 when Butler's biographer, Henry Festig Jones, published, with Francis's help, a pamphlet, *Charles Darwin and Samuel Butler: A Step Towards Reconciliation*, restoring Butler's reputation to some extent, if only posthumously.

Francis had given Jones access to relevant letters and he had both amended and then approved the final version of Jones's text. Francis admitted to Jones that at the very beginning of the dispute he and some of his brothers had disagreed with the advice of those, like Huxley, who urged Charles to ignore Butler; they had wanted Charles to publish a full account of what had happened, explaining that he had made a mistake, albeit an innocent one. No such explanation was ever made, so it can only be inferred that either a majority of the children, or at least the more forceful family members, had agreed with Huxley. Francis told Jones, 'I have often regretted that when the quarrel began I did not go to Butler and have it out *viva voce*. I also think I was mistaken in not publishing in *Life and Letters* a full account of the thing.'[49] Francis had felt bound by family loyalty. It is reported that, as early as 1894, he had let it be known that he hoped a third party would step in and heal the rift, something which at the time he felt unable to do himself.[50]

Both Charles and Francis lacked courage during the Butler Controversy; Charles to admit that he had made a negligent mistake; Francis to oppose his family in those first few years after Charles's death, when Butler was still alive and might have benefited from some sort of apology. Francis's instinct was to reject anything that might tarnish his father's image. In the case of the danger posed by Butler, this was in spite of his having some sympathy with Butler's arguments about evolution.

As much as ten years before his death, Butler himself could have found evidence of that sympathy if he had read a paper given by Francis to the Cardiff meeting of the BAAS (1891). Francis's subject was, 'On the Artificial production of Rhythm in Plants'. The paper was later published in the *Annals of Botany*[51] and it contains the sentence, 'This repeating power may be that fundamental property of living matter which stretches from inheritance on one side to memory on the other'.[52] Francis voiced this somewhat obscure proposition again several times over the following years, his thoughts on the matter, no doubt, maturing with the passing years. However, it was not until 1908, six years after Butler's death, that he revealed fully the rather shocking thinking behind his proposition. The occasion was his Inaugural Address given as President of the BAAS's Dublin meeting.[53]

His opening words sought to justify the direction of his thought by referring to two unimpeachable authorities, his father and Julius von Sachs,

> In his book on 'The Power of Movement in Plants' (1880) my father wrote that 'it is impossible not to be struck with the resemblance between the foregoing movements of plants and many of the actions performed unconsciously by the lower animals'. In the previous year Sachs had in like manner called attention to the essential resemblance between the irritability of plants and animals ... the fact that plants must be classed with animals as regard their manner of reaction to stimuli has now become almost a commonplace of physiology.[54]

He went on to propose that stimuli were not momentary in effect but left a trace of themselves on the organism and this was the physical basis 'of the phenomenon grouped under memory in its widest sense'. The zoologist, H. S. Jennings, had recently proposed – in *The Behaviour of Lower Organisms* (1904) – that in lower organisms, just as in man, behaviour is modified by their physiological condition and the reactions they have performed. Francis ventured 'to believe that this is true of plants as well as animals'.[55]

Adopting a term from the widely discussed writings of Richard Semon, Francis spoke of this form of memory as a 'mnemic' factor in the life of plants. Thus, plants whose leaves show sleep movements under normal day/night regimes continue to show leaf movements after the plant is moved into continuous darkness. While plants had no central nervous system, he argued, their complex system of nuclei did have 'some of the qualities of nerve cells, while intercommunicating protoplasmic threads may play the part of nerves'.[56] (The 'threads' were, in fact, the plasmodesmata discovered by Walter Gardiner; see Chapter 8).

Judged by the state of knowledge of the day, where biologists were struggling to place in a coherent framework the welter of new facts with which they were faced, and where the extent of any commonality between plants and animals was as controversial as it was uncertain, Francis's arguments thus far appear reasonable, although misguided. But he then went beyond the point where he could adduce evidence to support his arguments. He went beyond the point at which he could any longer be easily pardoned.

Firstly, in a remark that contained echoes of great grandfather Erasmus's 'sensorium' (see Chapter 4), he asserted, 'It is impossible to know whether or not plants are conscious; but it is consistent with the doctrine of continuity that in all living things there is something psychic, and if we accept this ... we must believe that in plants there exists a faint copy of what we know as consciousness in ourselves'.[57]

Secondly, he progressed from considering there being a mnemic quality in movements to there being a mnemic quality in morphological changes.

> The development of the individual from the germ-cell takes place by a series of stages of cell-division and growth, each stage apparently serving as a stimulus to the next, each unit following its predecessor like the movements linked together in an habitual action performed by an animal.
>
> My view is that the rhythm of ontogeny is actually and literally a habit.[58]

The ontogenetic rhythm had, he thought, both a fixed component, most readily seen in the early stages of development, and a variable component, most readily seen in the later stages. Pointing out that in *The Origin of Species* his father wrote, 'on the view that species are only strongly marked and fixed varieties, we might expect often to find them still continuing to vary in those parts of their structure which have varied within a moderately recent period',[59] Francis concluded

> Evolution ... depends on a change in the ontogenetic rhythm ... [for] if this rhythm is absolutely fixed, a species can never give rise to varieties.
>
> ... a beech tree may be made to develop different forms of leaves by exposing it to sunshine or to shade. The ontogeny is different in the two cases ... in some orchids the assimilating roots take on a flattened form when exposed to sunlight, but in others this morphological change has become automatic, and occurs even in darkness.[60]

Citing Ewald Hering, whose views were adopted by Butler in his book *Unconscious Memory* (1880), Francis proposed that unconscious memory resides in the germ cells. These cells, he said, had 'the power of retaining the residual effects of former stimuli and of giving forth or reproducing under certain conditions an echo of the original stimulus'. 'The mnemic theory of development depends on the possibility of what is known as somatic inheritance or *the inheritance of acquired characters*' [author's italics].[61] Francis had aligned himself with Butler and with Lamark. The momentum of his arguments had carried him onwards to a point where he could be regarded as being in an anti-Darwinian camp. In mitigation, it should be remembered that even Charles, when unusually he concerned himself with the mechanism of inheritance and pangenesis (see Chapter 5), allowed that messages from environmentally influenced somatic cells could be passed between generations.[62]

*****

To return for the last time to Ellen's death: the sale of Wychfield and the move to London were part of a plan devised by Francis not just to support his daughter but also to construct for himself a new and different life. Resigning his Readership in Cambridge – where many of his old students organized a leaving party at which he was presented with an address, and the Botany School was presented with a portrait of Francis by his friend Rothenstein – he was by a happy coincidence able to meet the current need of the Royal Society for a new Foreign Secretary.[63] Possibly wanting to avoid the unoccupied time that might allow him to think too deeply about his personal problems, he had swapped old responsibilities for new ones. He had probably realized, however, that, although the role of the Foreign Secretary was an onerous one, he could do the work at his own pace and, largely, at times of his own choosing. Francis filled the position with distinction from 1903 to 1909 and, at least partly in recognition, he was made the Vice-President of the Royal Society in 1907–8. The presidency of the BAAS in 1908 was not just another personal honour but an honour for the whole of botany, for it was the first time a botanist had been elected to the post since Joseph Hooker in 1868.

While managing to keep a modicum of research going in Cambridge – research which led to one of his most important papers, that with Pertz on estimating stomatal apertures (see Chapter 8) – a significant part of his later years was taken up once more with his father's affairs. He arranged for his father's scientific library to be housed in the Botany School in Cambridge, expressing his intention that it should be bequeathed to the University. A continuing worry for Francis and his older brothers, William and George, was the fate of Down House. The family made little use of it following Emma's death in 1896 and, after a few years, they decided to lease it to tenants. Thus, it was from1907 to 1922 in the charge of Olive Willis who used it as a school for girls. Towards the end of this period and, increasingly worried by the declining condition of the house, a number of eminent men called for it to be protected as a national monument but still little happened and more tenants came and went. William, George and Francis were all dead by 1929 when the BAAS finally stepped in to rescue Down, opening it to the public for the first time. Problems remained however for the long term maintenance of the house was too costly for the BAAS, as it was also for its later stewards, the Royal College of Surgeons of England, and the Natural History Museum. It was not until 1996 that the fate of the house was secured when English Heritage finally purchased and fully restored the house.[64]

To mark the centenary of Charles Darwin's birth, major celebrations were held in 1909. Cambridge University and the Darwin family collaborated in their planning. Francis was especially heavily involved. He contributed a chapter on 'Darwin's Work on the Movements of Plants' to a commemorative volume, *Darwin and Modern Science: Essays in Commemoration of the Centenary of the Birth of Charles Darwin,* edited by Albert Seward and published by Cambridge University Press. He edited and wrote a short introduction for a booklet given by the Syndics of the Press to everyone attending the celebrations, held in Cambridge, in June. Entitled, *The Foundations of the Origin of Species, a Sketch Written in 1842*, it was based on a long-lost manuscript of Charles's, only discovered at the back of a cupboard in Down House after Emma's death. Essentially a sketch of his ideas at the time it was written, Charles had mentioned it in his *Autobiography* as 'a very brief abstract of my theory'.[65]

Francis received yet another honour during the celebrations. The university conferred honorary degrees upon many of its invited guests to mark their distinction in the sciences. Francis was the only Englishman to receive such an honour, being awarded an Honorary Doctorate of Science. On 23 June a grand banquet was held in Trinity College at which the Rt. Hon. Arthur J Balfour, the ex-Prime Minister (and tennis partner of George Darwin when they were students), gave a speech reflecting upon Charles's achievements and his worldwide standing. William spoke for the family, by all accounts giving the best speech of the evening. Next day there was a huge garden party collectively organized by the Darwin family for their various house guests and friends. The spirit of the occasion is charmingly captured by Will Rothenstein's description of his fellow guests in Francis's house:

> Mrs Huxley, Thomas Huxley's widow, and Sir Joseph Hooker, both well over eighty, were staying with the Darwins. Were old world manners and charm, I wondered, more common in the past, or do they come with mature years? Mrs Huxley certainly had them, with a surprisingly alert mind. It was touching to see old Sir Joseph Hooker with Francis Darwin's little grand-child in his arms. I thought what a wide period would be covered if the infant lived to the scientist's great age.[66]

One celebration was shortly followed by another; on 1 July, Frances was married to Francis Cornford. At the age of sixty-one, Francis Darwin was freed to live his life at a pace befitting a sexuagenarian. Travel, music, bird watching and his favourite authors, Jane Austen and Charles Dickens, took more and more of his time. An especially happy year for him was 1913, for he was knighted and he married again, for the third time. His bride, Florence, was the widow of the Cambridge historian, F. W. Maitland. Described by Gwen Raverat, as 'strange and beautiful',[67] little is known about Florence, except that she and Francis settled in a large house at Brookthorpe, in the Gloucestershire Cotswolds. The fact that the Rothensteins lived at Iles Farm in nearby Oakridge suggests that, like her new husband, Florence must have enjoyed their company. Indeed, Rothenstein's diaries show the families frequently visited each other. At Iles Farm, the Darwins mixed with the Rothensteins' friends, including many leading figures from the world of the arts, such as Max Beerbohm, Augustus John and A. E. Houseman, and they were taken to village events. After enjoying an amateur dramatic performance at Oakridge, Lady Darwin was so delighted that she wrote a play, *The Briary Bush*, for the villagers. Their performance of it so pleased her that she was moved to tears.[68]

At peace in rural Gloucestershire, Francis compiled two small books, *Rustic Sounds* (1917) and *Springtime and Other Essays* (1920). Each contains a mixture of essays he had already published, such as 'Reminscences' of his father's life, the text of some talks he had given, for example, 'The Teaching of Science' (an address to Birkbeck College in 1913), and some new material reflecting his lifelong interest in early musical instruments and country dances. From these small books there emerges a unique picture of the author, his family, his science and his hobbies – not to mention his opinions. Chapters such as, 'A Lane in the Cotswolds' and ' The Traditional Names of English Plants', capture the spirit of tranquil retirement deep in the English countryside, far from the World War I and its aftermath.

As ever in Francis's life sadness was not far away, for in 1920 Florence died. His evening of happiness was at an end. Now in his seventies he pulled together what remained of his life and returned to his dear, familiar Cambridge where he tried in vain to pick up the vestiges of his old life. In truth he cut a lonely and pathetic figure in the eyes of younger men, like Harry Godwin (later head of the Botany School).

> In the early 1920s Francis Darwin represented for us a strong link with the pre-war period. Though retired, he came into the Botany School with fair regularity, a tall and impressive, though aging man, wearing a long snuff-coloured cloak and wide flat-brimmed hat. Plant material from the Botanic Garden was set out in his room by the senior laboratory assistant 'Henry' Elborn, together with simple apparatus devised some years earlier by Francis Darwin, who now busied himself with observations that were shortly written up for publication. It was however pathetically clear that the ageing scientist was in fact merely repeating work he had already done: it had indeed appeared in scientific journals and it was now the responsibility of the resourceful 'Henry' to secure the almost finished manuscript before it could be posted to an unsuspecting editor. This accomplished on one pretext or another, a new series of trials soon replaced the earlier one. It all amounted to a kindly service to a respected teacher now losing his touch.[69]

Francis died on 19 September 1925. His funeral service was held in held in the chapel of Christ's College before he was buried in the cemetery at Huntingdon Road. *The Times*, of Monday 21 September, recorded that among the University and scientific dignitaries in attendance were Dr Ralph Vaughan Williams (composer and member of the extended Darwin family), Major General Sir Richard Ruck (schoolfriend and brother of Amy), W. Rothenstein, F. F. Blackman, J. M. Keynes (friend, Nobel Prize-winning economist and brother-in-law of his niece Margaret), and Professor Sir Rowland Biffen (colleague and pioneer crop breeder).

10 FORTUNE'S FAVOURITES?

> Fortune, we are told, is a blind and fickle foster-mother, who showers her gifts at random upon her nurslings. But we do her a grave injustice if we believe such an accusation. Trace a man's career from his cradle to his grave and mark how Fortune has treated him. You will find that when he is once dead she can for the most part be vindicated from the charge of any but very superficial fickleness. Her blindness is the merest fable; she can espy her favourites long before they are born.[1]

*****

The advantages of wealth, social position and personal connections may have formed the springboard from which the scientific successes of the Darwins were launched, and in this respect they may have been fortunate, but the Darwins were not unique. Even within the confines of this narrative parallels to their lives can be found. Thus, Hales like Erasmus was free to pursue botany as a hobby in time left over from his secure, comfortable professional life. Thus, Dutrochet like Charles was able to lead the life of a wealthy country-house based gentleman scientist. Such stories were repeated many times over in the eighteenth and nineteenth century across Europe, and even men like Priestley and Ingen-Housz could point to some good fortune for they were able to find themselves wealthy patrons and protectors.

What was exceptional about the Darwins was that their wealth, social position and personal connections were assiduously nurtured (see Chapter 2), so that, as generations passed, Fortune was more and more generous in the gifts she bestowed upon her nurslings. What was unique about the Darwins was that different generations of one family contributed so much to the advance of one science. So, it is appropriate to ask, what were the *scientific* gifts that were passed between generations? How fortunate was Charles to have Erasmus as his grandfather and, in his turn, Francis to have sprung from Charles and Erasmus?

In earlier biographies of Erasmus and Charles, the men have been treated separately and their botany has taken only a small part; no biography has been written of Francis. The individual contribution of each man to the history of botany has been set out in previous chapters here, so now, for the first time, it is possible to ask, where their botany is concerned, how much did they owe to each other? And what did they achieve *collectively*? Although this chapter addresses both questions, a complete answer to the second must await the perspective provided by the final chapter – which considers how their legacy was used by later generations of botanists.

Francis Darwin is particularly interesting for what he represents historically. He was one of the new breed of professional, laboratory-based scientists, yet he was blessed with the advantages of personal wealth and family connections. He was a pointer to the future but a reminder of botany's past, some habits of which persisted well into the twentieth century.

It was Francis who stood to gain most from the other Darwins, simply by virtue of his being last in the line, so this chapter will begin by asking, was he always destined to be a winner, smiled upon throughout his life by Butler's Fortune? Was it fair that a contemporary commentator should make remark, '[the sons of Charles were] the friends of eminent men but not eminent men in their own right'?[2]

*****

The sons of Charles were indisputably eminent. Three were knighted and elected to Fellowships of the Royal Society, Francis the botanist, George the mathematician, and Horace the scientific instrument maker.[3] A fourth son, Leonard, served as President of the Royal Geographic Society. (From Dr Erasmus Darwin's election to the Royal Society in 1761, through to the death in 1962 of Sir George Darwin's son, Charles G., the family celebrated a continuous line of Fellows lasting 201 years – an unique dynasty.) But were they, or, in this case, was Francis eminent in his own right?

By the time Francis was born, much would have been expected of a young man bearing the Darwin name, so in one respect it was a burden, but for a man with a talent for science any disadvantage must have been greatly outweighed by the opportunities that accrued from being a Darwin. Charles was an exalted figure in British scientific society in his lifetime and for almost a hundred years from the mid-nineteenth century Darwins were firmly embedded in what Noel Annan has called the 'intellectual aristocracy'.[4] Darwins were related to, or were friends of, the most influential men and women in society. Among this elite, the Darwins encouraged and supported each other.

Beginning with Charles's marriage to Emma Wedgewood (her family similarly made exceptionally good connections through marriages), the astonishing web of family connections traced by Annan is too extensive to repeat here but two examples, both involving Charles's sons, give its flavour. The first involves

Francis's third wife, Florence. She was both the sister of Herbert Fisher – from 1916 President of the Board of Education in Prime Minister Lloyd George's government – and sister-in-law to the composer Ralph Vaughan Williams. The second involves George's daughter, Margaret. She married the eminent surgeon, (Sir) Geoffrey Keynes, who was the brother of Nobel Prize winning economist (Lord) John Maynard Keynes.

Apart from their marital connections, the Darwins were exceptionally skilled at networking. Erasmus had had his circle of Lunaticks, and, as seen in Chapters 2 and 5, Charles depended on his own circle of friends and correspondents, such as Hooker and Gray, for help and advice, while his brother (Ras) and wife introduced him to such luminaries as Tennyson, Ruskin and Munby. Charles's children numbered amongst their friends the most powerful and influential men and women of their day. For example, Arthur J. Balfour, Prime Minister from 1902 to 1905, was an old friend of George Darwin, their friendship having been forged when as undergraduates they competed for a place in the Cambridge tennis team to play against Oxford.[5] Balfour's sister, Eleanor, married Henry Sidgwick. Henry, as seen in Chapter 9, did more than anyone else to create Newnham College and, after his death, Eleanor was Principal of the College. One of Henry's cousins was Francis's second wife, Ellen, an early Newnham lecturer.

After the Darwin family quit Downe, following Charles's death, Cambridge became their powerbase. Even in the early years of the twentieth century Cambridge remained a small university town. Set in the heart of flat, rural East Anglia – what Frances Cornford called, 'Our sober, fruitful, unemphatic land'[6] – it had been connected to London by railway since 1845, and to Oxford since 1862, but an aura of geographical isolation still hung around it. Francis Darwin's son-in-law, Francis Cornford, summed it up:

> The resident dons constituted a political society which was small enough for everyone to know, or to know of, everyone else. Cambridge had about it something of the cosiness (and also, no doubt, the claustrophobia) of village life, a quality greatly enhanced by the connexions of kin that also bound the dons together.[7]

Their wives formed cliques where gossip and intrigue flourished. Among the most exclusive, and one riddled with Darwin women, was the Ladies' Dining Society. Its members were Mrs Creighton, Mrs Arthur Verrall [formerly Miss Merrifield, ex-Newnham], Mrs Arthur Lyttelton, Mrs Sidgwick [second Principal of Newnham], Mrs James Ward [Mary Martin, ex-Nuneham], Mrs Francis Darwin [Ellen Crofts, ex-Newnham], Baroness von Hugel, Lady Horace Darwin, Lady George Darwin, Mrs Prothero, Lady Jebb, and Mrs Paley Marshall [the first female lecturer appointed by Newnham].[8] And, if the Darwins' pre-eminence in academic Cambridge society were not enough, Leonard made sure the family name was at the forefront of the city's society when he was elected its mayor. Neither gown nor town could ignore the Darwins.[9]

In his Obituary of Francis, Albert Seward remarked, 'His hours of leisure were not curtailed through the necessity of adding to income by any form of drudgery'.[10]

Francis – like his brothers – had no need to pursue a career, to expose himself to the possibility of criticism and condemnation for not matching up to his father (or his great grandfather). But he took the risk. He researched, published and stood by his own conclusions and opinions. He was a pioneer in the study of plant movements and an early champion of the practical teaching of plant physiology, making major contributions in both areas *after* his father's death. He played a full part in the conduct and administration of science nationally, for he was a leading figure in the BAAS and a distinguished Foreign Secretary of the Royal Society. And his editing of the life and letters of one of the most famous men in the world won him critical acclaim.

The Darwin name must have helped Francis – it was through Charles's manipulations that his first botanical paper was published – for then, as now, 'who you know' was important in the not-totally objective world of science. But Francis achieved more than enough *in his own right* to deserve to be called eminent.

*****

> The external crust of the earth, as far as it has been exposed to our view in mines or mountains, countenances this opinion; since these have evidently for the most part had their origin from the shells of fishes, the decomposition of vegetables, and the recrements of other animal materials, and must therefore have been formed progressively from small beginnings. There are likewise some apparently useless or incomplete appendages to plants and animals, which seem to show they have gradually undergone changes from their original state; such as stamens without anthers, and styles without stigmas ...[11]

Clearly Erasmus had anticipated by half a century key arguments in his grandson's great evolutionary theory. His reason and foresight can only be marvelled at when, in *The Temple of Nature*, he suggested simply and unequivocally that life arose from a single common ancestor, forming 'one living filament'.

> Organic life beneath the shoreless waves
> Was born and nurs'd in ocean's pearly caves;
> First forms minute, unseen by spheric glass,
> Move on the mud, or pierce the watery mass;
> These, successive generations bloom,
> New powers acquire and larger limbs assume;
> Whence countless groups of vegetation spring,
> And breathing realms of fin and feet and wing.[12]

And, of the struggle between species, which involved plants as much as animals, he wrote:

> Herb, shrub, and tree, with strong motions rise
> For light and air, and battle in the skies;
> Whose roots diverging with opposing toil
> Contend below for moisture and for soil.[13]

The present book has not been concerned with Charles's evolutionary theories per se, but rather it has explored the evidence he garnered from plant biology to support those theories. However, where Charles's debt to Erasmus is concerned, the origins of his thoughts about the evolution of living organisms is relevant, although it remains puzzling. Extracts from *The Economy of Vegetation* and *The Temple of Nature* clearly show that Erasmus had anticipated key arguments in grandson's great theory. Yet Erasmus is not mentioned in the first edition of the *Origin of Species* and while those references to 'my theory', that are so common in the first edition, gradually disappear in later editions, mention of Erasmus is still restricted to a note in small print saying that he 'anticipated the views and erroneous grounds of opinion of Lamark'.[14] Desmond King-Hele, who has devoted many years to researching and chronicling Erasmus's life, generously concludes that a person so transparently honest as Charles could only have been unintentionally unjust to his grandfather.[15] King-Hele goes on to suggest, however, that Charles was anxious not to be associated with the dated and discredited writings of his grandfather on evolution. Later in life, when Charles was seventy years old, he wrote his excellent *Life of Erasmus Darwin* but even what was at face value a tribute – it might have been thought of an as act of contrition for earlier oversight – reveals an enduring ambivalence towards Erasmus. For one thing, Charles was quite happy that what he had written should be published, together with a far shorter essay by Ernst Krause (see Chapter 9), in a book with Krause's name, rather than Charles's, on its title page. Krause has, therefore, been widely credited in bibliographic systems with authorship of the whole. Possibly it was Charles's intention to praise his grandfather by proxy, so he may have been content that the most glowing tributes should appear in Krause' section of the book. However, that seems unlikely for it now appears that that even what was printed under Charles's name had been heavily censored at the proof stage by his daughter, Henrietta. She had cut almost one eighth of the text, removing not merely any sections that might possibly have been offensive to churchmen but also many of her father's comments that were complimentary to Erasmus, such as, 'he anticipated many new and now admitted scientific truths'.[16] (In a similar way she later forced Francis's hand when he was editing *Life and Letters*. Henrietta was zealously and needlessly over-protective of her father's reputation; see Chapter 9).

What else is known of Charles's attitude to Erasmus is his admission that when he was about seventeen years old he read and 'greatly admired' Erasmus's *Zoonomia*, although, he said, it produced no effect on him.[17] He said further that only years later, and after he had already become convinced of the reality of natural evolution, did he read *Zoonomia* again, this time being 'much disappointed, the proportion of speculation being so large to the facts given'.[18] This may have been his honest belief but there are in the *Origin of Species* what King-Hele calls 'subconscious echoes of Erasmus',[19] for example, where Erasmus had written:

> the strongest and most active [male] animal should propagate the species, which should thence become improved[20]

Charles wrote:

> The most vigorous males, those which are best fitted for their places in nature, will leave most progeny[21]

There may have been other lessons learned unconsciously. For Erasmus, sexual reproduction was superior to asexual because it led to variety amongst progeny, a circumstance 'employed to great advantage by skilful gardeners'.[22] Charles similarly saw that sexual reproduction and outbreeding were the keys to generating the variation that was the raw material upon which natural selection acted, but he went further. He demanded to know whether there were limits to variability and never stopped quizzing gardeners about the weird and wonderful plants that they could produce by manipulating cross pollinations.

Turning away from evolutionary theory and towards Charles's practical, botanical studies, what unquestionably connects Charles to Erasmus is their mutual fascination with insectivorous plants, in particular, and with plant movements in general. There is clear evidence that Charles often focussed his attention on the same plants that had featured in his grandfather's writings. Thus, among the insectivores, Erasmus picked out *Drosera* and the Venus Flytrap, describing their form and speculating upon their habits. The same two plants received special attention from Charles, as he examined a lengthy list of organic and inorganic materials that might or might not cause the plant to respond (see Chapter 6). Charles told Lyell, I care more about *Drosera* than the origin of all the species in the world'.[23] And it was the Venus Flytrap that Charles persuaded Burdon-Sanderson to use in his experiments examining the transmission of stimuli. Few detailed drawings of plants are included in *The Botanic Garden* but two are of insectivorous plants and one is of a climbing plant.

At the very beginning of Charles's interest in plant movements, his attention was taken by *Hedysarum gyrans*, none other than the 'Fair Chunda' of Erasmus's *Loves of Plants*: chapter 3. As early as 1855, that is eighteen years before the beginning of the period when he concentrated on plant movements, Charles was writing to Hooker:

> I thank you much for *Hedysarum*: I do hope it is not very precious, for, as I told you, it is for probably a most foolish purpose. I read somewhere that no plant closes its leaves so promptly in darkness, and I want to cover it up daily for half an hour, and see if I can *teach it* to close by itself, or more easily than at first in darkness.[24]

His request for plants was repeated in 1862.

If Charles had profited by carrying away the message of Erasmus's writings without being encumbered by their often fanciful details, then there was one area where he suffered by not paying enough attention to detail. The 'Geographical Distribution of Plants' formed a whole chapter in the first edition of the *Origin of Species*, for it was important for Charles to prove that new species of plants (and animals) could spread rapidly over large parts of the earth. In this context Charles tested the ability of seeds of many species to withstand immersion in sea water, concluding that, 'seeds of about 10/100 plants of a flora, after having been dried, could be floated across a space of sea 900 miles in width, and would then germinate'.[25] If he had paid better attention to Erasmus he would have remembered a reference to the voyages of seeds of *Cassia*, a plant which grows in America but whose seeds are regularly washed up on the coast of Norway:

> Soft breathes the gale, the current gently moves
> And bears to Norway's coasts her infant-loves[26]

Erasmus lists in his footnotes several other American fruits, such as the cashew nut, bottlegourd, logwood tree and cocoa-nut, that were known by Dr Tonning regularly to cross the 3,000-mile wide Atlantic ocean. Another reputable source, Sir Hans Sloane, had, Erasmus wrote, recorded four species of West Indian plant whose seeds were regularly washed up on the shores of the British Isles. Charles had missed good evidence of viable seeds making journeys much longer than he had dared to suggest.

If Charles's debt to Erasmus was greater than he cared to acknowledge then it is only fair to point out that Erasmus too has not been free from the charge of borrowing from others. In this case the offended party was Anna Seward, daughter of the Canon of Lichfield Cathedral, whose considerable poetic talents Erasmus had encouraged. Erasmus cultivated the friendship of the young woman and introduced her to several members of the future Lunar Society, with whom she forged a variety of relationships. But it was the one with Erasmus that was to prove most tempestuous for what was once friendship – or possibly more – changed into bitter acrimony as in her *Memoirs of the Life of Dr Darwin* (1804), she charged him with stealing her ideas. According to Anna, she was one day sitting alone in Erasmus's newly created botanic garden, her thoughts absorbed by the variety and beauty of the flowers, when she had the idea of writing verses in which plants were represented as people; in effect she was reversing Ovid's method of representing people as plants. She and Erasmus could collaborate: she would write verses while he would add scientific notes. Erasmus, not unexpectedly, recalled that the idea of personifying plants was his own, although he was reluctant, at first, to jeopardise his serious medical credentials by publishing poetry. Whoever was its progenitor, the idea was realised in Anna's *Verses Written in Dr Darwin's Botanic Garden*, a short poem that Erasmus had arranged to be published in the *Gentleman's Magazine* in 1783. Instead of being grateful, Anna was justifiably annoyed because, although the verses appeared under her name, she had not given her permission for either their publication or the alteration by Erasmus of their first lines. Further injury was caused when the verses, modified again, were used as the opening lines of *The Botanic Garden*. Again Erasmus did not consult Anna, although this time he did acknowledge her contribution.[27] In fairness to Erasmus, he had, apart from these few lines, written all the poetry and the notes himself.

*****

Francis's botany contained echoes of his great-grandfather. His inspiration for the horn-hygrometer that he devised for measuring transpiration and, thereby, stomatal apertures could well have been drawn from Erasmus's *Loves of the Plants*:

> The seed vessel [of Impatiens or Touch-me-not] consists of one cell with five division: each of these, when the seed is ripe, on being touched, suddenly folds itself into a spiral form, leaps from the stalk, and disperses the seed to a great distance by its elasticity. The capsule of the geranium and the beard of wild oats are twisted for a similar purpose, and dislodge their seeds on wet days, when the ground is best fitted to receive them. Hence one of these, with its adhering capsule or beard fixed on a stand, serves the purpose of an hygrometer, twisting itself more or less according to the moisture of the air.[28]

And the pores of leaves, later known as stomata, fascinated Erasmus (see Chapter 4), just as they did Francis.

The way that Francis with his limited knowledge of botany, but his recent experience of an animal physiology laboratory, became in 1874 his father's secretary, then his assistant, and finally his collaborator, has already been told (Chapter 7). From Charles he learned in that period, if he had not already learned during his childhood at Downe, the importance of careful, accurate observation. In this respect Charles was the perfect role model, his studies of whatever group of organisms was his focus were as exhaustive as they were exhausting. Typically he liked to work from a blank sheet, without too many preconceived ideas about a plant (a style that is reflected in his botanical writing). Having chosen his main species for study, he would repeat his observations on this, or a very few, species many times. His next step was to test his major conclusions on a range of other species, often, it seems, the more distantly related the better. For example, when studying nyctitropism for the *Power of Movement in Plants*, he and Francis first made detailed measurements on seven genera, *Cassia* (2 species), *Lotus*, *Melilotus* (5 species), *Marsilea*, *Mimosa*, *Oxalis* (5 species), and *Trifolium* (2 species). Next, they made less exact but nevertheless equally careful observations on representatives of no less than fifteen of the Tribes of the Leguminosae, including, in Tribe 15, plants native to Australia, India and South America. Francis had much to learn from his father about method, diligence and persistence, and, at the beginning of their collaboration, Charles probably knew more plants than did Francis although, in contrast, in day-to-day technical matters Francis had little to learn.

Charles might be thought a model laboratory scientist, in spite of working in a country house, but, in some ways, he seems to have been curiously unsure of himself in the laboratory, and sometimes unquestioning. Thus, when studying insectivorous plants, he wanted to find how small a particle would cause the tentacle of *Drosera* to inflect. He writes, 'Measured lengths of a narrow strip of blotting paper, of fine cotton thread, and of a woman's hair, were carefully weighed for me by Mr Trentham Reeks, in an excellent balance, in the laboratory at Jermyn Street'. 'Short bits were then cut off and measured by a micrometer, so that their weight could be easily calculated'.[29] What stands out is that neither was he prepared to invest in a precision balance himself, nor did he consider how the weight of such small samples might be affected by the humidity of the atmosphere and, therefore, the different amounts of water they might hold in his room and in Mr Reeks's laboratory. Similarly, ammonium phosphate was weighed for him by 'a chemist with an excellent balance'.[30] In his studies of the survival of seeds, artificial sea water was made up for him, while in another investigation no less a person than Sir Edward Frankland supplied him with distilled water. The ugly question arises, was Charles unsure of himself, or was he so used to having servants to attend to his every domestic need that he didn't expect to have to do anything tedious in the laboratory? Those best placed to answer, such as Francis and Henrietta, would certainly never have divulged such weaknesses, if they had existed, so the question must remain unanswered, though even the loyal Francis does hint in *Life and Letters* that his father's imprecise measurements and lack of modern apparatus were frustrating.

Remember, in Sachs's laboratory, Francis had seen the precision balances, smoked recording drums and amplifiers that were already familiar to him from Foster's laboratory. He knew such sophisticated apparatus could be applied to plants just as usefully as to animals. The expensive Hartaack microscope listed in Charles's effects at his death was probably chosen and brought back to Down House by Francis,[31] for after his father's death he wrote:

> It strikes us nowadays as extraordinary that he should have had no compound microscope when he went [*sic*] his 'Beagle' voyage; but in this he followed the advice of Robert Brown, who was an authority on such matters.[32]

Also,

> I have always felt it to be a curious fact, that he who has altered the face of Biological Science, and is in this respect the chief of the moderns, should have written and worked in so essentially a non-modern spirit and manner.[33]

> If any one had looked at his tools, &c. lying on the table, he would have been struck by an air of simpleness, make-shift, and oddness.[34]

Those oddities included a seven-foot ruler calibrated by the village carpenter and, in poor imitation of Sachs, some small square pieces of glass, coated with lamp black, with which he measured the circumnutations of root tips (Chapter 6). Francis adds:

> He had a chemical balance which dated from the days when he worked at chemistry with his brother Erasmus.[35]

> Measurements of capacity were made with an apothecary's measuring glass – I remember well its rough look and bad graduation.[36]

> Considering how naturally tidy & meticulous he was in essential things it is curious that he put up with so many makeshifts, it did not occur to him to have anything made for a special purpose, but fitted it up as well as he could, he would not have for instance a box made of the desired shape & stained black inside, but he would hunt up something like what he wanted & get it darkened inside with shoe blacking; he did not care to have glass covers for tumblers in which he germinated seeds made but used broken bits of irregular shape with perhaps a narrow angle sticking uselessly out on one side. But so much of his experimenting before the Power of Movement had been of the simplest kind.[37]

Inclusion of 'before' is, for Francis, quite outspoken and highly significant. It is telling evidence of the modernizing influence that Francis had upon his father and his part in the practical work reported in *The Power of Movement*.

Outsiders may not have been privy to Charles's old-fashioned methods but they could certainly recognise the peculiarly old fashioned surroundings in which he worked. While his sons were deciding what to do with Down House after his death (see Chapter 9), the Royal Society's 'Committee for Conducting

Statistical Inquiries into the Measurable Characteristics of Plants and Animals', which included among its members both Francis Darwin and William Bateson, and which was chaired by Charles's cousin, Francis Galton, suggested that the house and gardens should be acquired by the Society for use as a research station for breeding experiments. On this occasion, and also when the idea was revived in slightly different form in 1899, the idea was vigorously opposed by Bateson who, among others, pointed to Down's unsuitability, lacking as it did both proper laboratory facilities and a well stocked library.[38]

In one way Francis did manage to modernise his father and that was in his writing. In its succinct style and quantitative results, The *Power of Movement* bears a strong resemblance to Francis's style as seen in his 1878 paper on insectivorous plants, published in *Nature*.[39] Charles's later botanical publications are also much better referenced than his earlier ones, in particular showing an improved awareness of work published in German journals. This difference too must be attributed to Francis's influence for, as he recounted in his 'Reminiscences of my Father's Everyday Life',

> Much of [my father's] scientific reading was in German, and this was a serious labour to him. In reading a book after him, I was often struck at seeing, from the pencil marks made each day where he left off, how little he could read at a time. He used to call German the 'Verdammte', pronounced as if in English. He was especially indignant with Germans, because he was convinced that they could write simply if they chose. He learnt German by 'hammering away' with a dictionary. He said he read sentences many times over and at last the meaning occurred to him. He always pronounced German as he would do English. His bad ear for vocal sounds made it impossible for him to perceive small differences in pronunciation.[40]

To summarise what Francis gave to his father, it is unnecessary to look beyond Francis's own words

> He was largely dependent on the work of others for the facts used in the evolutionary work, and despised himself for belonging to the 'blessed gang' of compilers. And he correspondingly rejoiced in the employment of his wonderful power of observation in the physiological problems which occupied so much of his later life.[41]

Perhaps the greatest legacy that Charles gave in return to Francis – and it is unimaginable that he would have given such a gift to anyone other than his son – was his experience concerning the movements of insectivorous and climbing plants. A continuity of ideas was Francis's inheritance, and he used it well.

*****

Charles and Francis's joint studies of the movements of shoot and root tips, particularly of the bending of oat coleoptiles, had ramifications of the first importance, as will be seen in the next chapter. There too it will be seen that Francis's studies of stomatal movements led directly to our modern understanding of the role of plants in combating climate change. Francis may not have excelled in disparate subjects, as did Erasmus, or achieved the universal fame of Charles, but, where he is concerned, the comment about the sons of Charles, that they were 'The friends of eminent men but not eminent men in their own right',[42] is grossly unfair. Undoubtedly favoured by fortune, he nevertheless achieved eminence in his own right. Where plant science is concerned, he did more than 'live in his father's shadow'. He grew away from it, in the process putting plant physiology in Britain on a sound basis.

Erasmus, Charles and Francis worked against fundamentally different backgrounds. The three entered at radically different times onto the stage on which the history of scientific botany was being played out. Each man was a product of his age, shaped by particular teachers, both formal and informal, by books read, and, of course, by friends and contemporaries with similar interests. It is not appropriate – it is even invidious – to try to separate or compare them. Indeed, simply because they were part of a close-knit family, their efforts were synergistic.

Each of the Darwins flirted with the possibility that plants are sentient beings – Erasmus proposing there was a brain-like sensorium in each leaf or bud, Charles, more cautiously, writing about 'diffuse nervous matter' in *Drosera* (chapter 6) and radicle tips acting like brains (chapter 7), and the elderly Francis proposing there might be a physical basis for stored memory (the mnemic hypothesis) in plants – but no one person or laboratory did more than the Darwins, *collectively*, to prove that plants are sensitive beings, responding continuously to changes in their biotic and abiotic environments. Far from being isolated, their work was wholly in the mainstream, as the final chapter will underline.

Above all, *collectively*, they made an unique contribution to the dawning of plant science, helping botany become independent, free of medicine, and taken forward by professional men and, ultimately, by professional women.

11 WHERE DID THE GREEN THREADS LEAD? THE BOTANICAL LEGACY

The world of professional scientists from which Francis departed when his career finally descended into the confusion of senility was drastically different from the amateur world of natural philosophers in which Erasmus involved himself 150 years earlier. Francis, Erasmus and Charles, with other members of the Darwin family in minor parts, had helped to change the old world and in doing so had helped define the origins of scientific botany. One family had made a quite unparalleled contribution to the development of a fledgling science. By 1925, the year Francis died, the young science was well established. To appreciate their legacy fully it is, however, necessary to follow the evolution of botany some way beyond 1925, given the caveat that the influence of the Darwins was inevitably diluted with each succeeding decade, and each generation of botanists. The most important threads, however, extend right up to the present day. The botanical legacy of the Darwins is the subject of this last chapter.

Their legacy was transmitted in several ways. The first of these is most difficult to pin down for it involves the spirit of the age; the way in which men thought about their world. Thus, Erasmus helped change the spirit of his age through his writings and by bringing together men who were thinkers and doers. Their interests were diverse but from each other they drew the strength to go forward, overcoming practical and personal set-backs. Practical enquiry, they increasingly recognized, could benefit man's wealth, as well as his health. Charles repositioned man in the world, placing him firmly *in* the natural world rather than above and apart from it. He challenged religions, the structure of society and, what is of particular relevance here, he gave altered direction to much of the science of the day. His evolutionary studies challenged biologists and geologists with countless new questions, which most were eager to tackle. Not everything about the Darwinian revolution was positive, however, for while *On the Origin of Species* proved an incomparable stimulus to scientific enquiry, concepts such as 'the survival of the fittest' offered to lazy enquirers a ready-made framework into which they could inappropriately and all too easily shoe-horn their results. Also, at least in botany, *On the Origin of Species* led to an over-emphasis on comparative anatomy and morphology, and a loss of focus on physiology with its potential for practical improvement, as will be seen later in this chapter.

Charles's influence on botany was realized in a second way, which is through his worldwide list of contacts and correspondents (see Chapter 5), whose own practical work he helped to shape with his suggestions and advice. Men like John Scott the head propagator at the Royal Botanic Garden, Edinburgh, who examined hundreds of *Primulas* under instruction from Charles,[1] may have in one sense have worked for him informally, but he never had students or collaborators in any formal sense. There was no handing on of a legacy from person-to-person, excepting, of course, to his son, Francis.

Francis's world, bridging the nineteenth and twentieth centuries, is more familiar to our twenty-first century eyes. In contrast to his father, he could pass his wisdom and knowledge directly to others because, working in a university environment, he taught students and had paid research assistants, plus various collaborations. Generous by nature, he used his own reputation, not to mention the Darwin name, to help others whether by finding money for their research, or communicating their papers to the Royal Society, or, with a positive report, persuading an editor to accept for publication the paper of a young colleague.

The third and most tangible way in which Charles and Francis affected the history of botany is through the methods of experimentation that they developed and, of course, their discoveries. Sometimes transmission relied on personal contacts, as in the case of Francis and F. F. Blackman, but more often than not their works were taken forward by strangers having no personal connections.

*****

What was the botanical legacy of the first of the trio, Erasmus? At a practical level, he made botanical terminology more exact and made plant taxonomy more natural. With his unique literary style, he made botany exciting, drawing his readers' attention to recent advances in the understanding of plant function and chemistry, while contributing some original practical observations of his own (see Chapter 4). At a more general level, and maybe even more importantly, he hijacked for his own purposes the new concept of 'Nature' moulded by the Romantics and Naturphilosophers. They had changed the common definition of 'Nature' for the better. To the earlier rationalist philosophers of the eighteenth-century Enlightenment, 'Nature' meant simply the inherent properties of a thing – its distinctive and unique features. To Rousseau, Hegel and Schelling, nature was more than this. It was the sum total of forces and powers at work in the universe. It was, above all, dynamic, subject to change and causing changes.[2] This was exactly the nature about which Erasmus wrote, albeit in a seductively romantic style. For him, earth, air and living organisms all interacted dynamically. Where the Romantics and Naturphilosophers scorned measurement and experiment as a means of understanding this new nature, however, Erasmus lauded the practical endeavours of Hales and Ingen-Housz in botany, of Priestley and Lavoisier in chemistry. These were the philosophers whose example the young enquirers of the nineteenth century should follow, he argued, if nature was to be understood.

*****

The general perception of plants, even that held by mature professional botanists such as Joseph Hooker, was invigorated by Charles's treatment of them in *On the Origin of Species*, *Orchids* and *Insectivorous Plants*. Hooker in 1862 spoke of 'your jolly dancing facts anent [in respect to] orchid life'[3] and to many it seemed as if Charles – ably assisted by Burdon Sanderson among others – was transposing the observable motion and vitality of animal physiology into the erstwhile static and dull life of plants. The study of plants was given new energy and direction.

One direction stimulated by *Insectivorous Plants* was the study of nutrition in general and the role of 'ferments' in particular. Charles had drawn botanists' attention to the nutrition of plants, recognising the need of insectivorous ones, growing in nutrient poor environments, to supplement their supply of nitrogen by acquiring animal proteins (see Chapter 6). Many botanists failed, however, to recognize that Charles's conclusions could be extrapolated to the case of other plants growing in different, but similarly nutrient-poor, environments, and that they had relevance more widely to the subject of plant nutrition. Not so Wilhelm Pfeffer, who was used to finding gems in Charles's botanical books (see Chapter 6). Pfeffer, and a few others of equally perceptive mind, immediately recognized the importance of the studies of insectivorous plants. Thanks to them and to Charles, plant nutrition rose to the top of the physiological agenda. In Germany, Pfeffer and his collaborators set themselves the task of understanding how germinating maize seeds mobilize the insoluble nutrients stored within their endosperm. In Britain, where the study of germination was more practically focussed, being concerned with the conversion of stored starch into metabolizable sugars during the malting process, brilliant work was carried out in the laboratories of the brewers, Worthington, under the direction of Horace T. Brown. By 1890 he and his colleague Morris[4] were able to describe in great detail how the seed's scutellum secreted diastase [an enzyme] which when it reached the endosperm digested, or solubilized to sugars, the insoluble starch reserves stored therein. He also discovered cytase – which later proved to be a cellulase enzyme – which broke down cells walls within the endosperm, thereby facilitating the attack of diastase on starch.

Starch storage and mobilization within the leaf, rather than the seed, continued to pose questions for botanists. In 1891, Vines, whom it may be remembered had been the first to obtain from an insectivorous plant an extract with digestive powers (see Chapter 6), described a diastase in the leaves of grasses. Its role, he thought, was to release, from starch, sugars that could be translocated to other parts of the plant. He was correct, but implicit in his thinking was an acceptance of Sachs's doctrine that starch was the first long-lived product of photosynthesis. In 1893, Brown and Morris published a lengthy paper,[5] based on the most detailed and accurate measurements. This landmark proved finally that sucrose (common sugar), not starch, is the first product of photosynthesis to accumulate. Sucrose does not build up to high concentrations, however, because it is converted to an insoluble store, starch, whenever its supply exceeds either its consumption in respiration or its translocation out of the leaf towards the growing points of the plant. Examining a much wider range of plants than Vines had done, Brown and Morris observed periodic rhythms in the activity of diastase, rhythms that broadly correlated with those of starch accumulation and utilization. (This was the same H. T. Brown who signally advanced understanding of CO_2 diffusion through stomata, as described in Chapter 8.)

Through Sydney Vines, the Darwins are connected in one more way with the early history of biochemistry. Vines, who continued his studies of digestive enzymes even after his move to Oxford, taught and inspired Joseph Reynolds Green. An early 'Lecturer in Vegetable Physiology' at Liverpool University, where he pioneered studies of plant biochemistry, Green later returned to his alma mater, Cambridge University, where he wrote his histories of botany.[6] He was, with Sachs and Pfeffer, largely responsible for the abandonment in the 1890s of the erroneous distinction between 'organized' ferments (within the protoplasm) and 'unorganized' ferments (extracellular cocktails, such as in the pitcher of *Nepenthes*). Removal of this distraction cleared the way for a more integrated approach to digestion and metabolism and, coincidentally, the more general adoption of the word 'enzyme'. Green clung, however, to the outdated belief that organisms exhibited metabolism in two distinct ways, by the (non-enzymic) mediation of protoplasm and by the action of enzymes.[7]

Charles declared that 'Making out' the structures of heterostylous primrose and flax flowers, and untangling the mysteries of legitimate and illegitimate crosses (see Chapter 5) had given him more satisfaction than anything else in his scientific life,[8] but he could hardly have guessed where his simple experiments would lead. Heterostyly is now known in twenty-eight different families of flowering plants and in many of these, such as in the genus *Primula*, there are examples where its genetic basis is well understood. In *Primula*, differences in style length, pollen size (thrum is larger than pin), and anther position are controlled by genes designated *G*, *P* and *A*, respectively. The three are tightly linked, on one chromosome, at what is known is known as the *S* locus. Pin plants are homozygous for the *s* allele (*ss*), and thrum plants heterozygous with a dominant *S* allele (*Ss*). Charles's observation that pin-thrum and thrum-pin crosses yielded equal numbers of pin and thrum progeny is exactly what would be expected from crosses between *ss* and *Ss* parents.[9]

In yet another example of Charles's insights into the lives of plants he had realized that, over and above any physical limitations on cross-pollination involving various morphs of the same species, there is in illegitimate crosses physiological incompatibility between the male pollen and receptive female tissues. Although he could manually transfer pollen to tissues that it would not normally reach, fertile offspring still did not result. The subject of self-incompatibility was largely ignored until the 1940s when his basic findings were confirmed and then greatly extended by R. A. Fisher and Kenneth Mather, Head of the Genetics Department at the John Innes Institute.

Debate still rages as to whether the initial step in the evolution of floral polymorphism was a response to the selection pressure for efficient cross-pollination, and self-incompatibility then followed, or, as has been more commonly assumed, genetic incompatibility arose first in monomorphic types, in response to the comparative un-fitness of self fertilized types, and the physical separation of male and female organs (herkogamy) came second.[10]

In this, as in many other instances, complete answers still elude us. But in this, as in many other instances, Charles had opened up an important area of study simply by asking the right questions.

*****

> I greatly profited from the thorough knowledge my father had of botany in general. He knew the entire plant physiological literature, having read every important paper ever published (and remembering its content) ... He was friendly with most botanists all over the world, and thus I came to know many of them.[11]

If the words 'plant physiological' were excluded, these statements could have been written by Francis Darwin. They were not. They were written instead by Frits Warmold Went. His father, F. A. F. C Went, was professor of botany and director of the Botanical Gardens at the University of Utrecht, in the Netherlands, where Frits was born in 1903.

Frits Went made many significant contributions to twentieth-century botany but none was more important than one he made while a postgraduate in his father's laboratory.[12] His starting point was the discovery by Charles and Francis Darwin (1880) that the tip of grass coleoptiles was necessary for phototropism but that bending takes place in a region below the tip (see Chapter 7). His epoch-making conclusions, published in 1928, led in a very few years to the isolation and chemical characterization of auxins, and subsequently of other families of plant hormones. From this narrow starting point, Went opened up a research area of seemingly limitless practical applications.

Little substantive had happened in the study of phototropism in the years between 1880 and 1928. Pfeffer had written at length about the subject but his view that stimulus and response were analogous to the pull of a trigger and the explosion of a gun was unhelpful. He did contribute, nevertheless, to the ultimate solution of the problem for both Peter Boysen-Jensen and Arpad Paal worked for periods in his laboratory.[13] It was the Danish plant physiologist, Boysen-Jensen, who in 1913 kick-started progress when he found evidence that the 'influence' passing between tip and response zone was a chemical. Its movement could be blocked by a physical barrier. When he inserted a mica sheet halfway across a unilaterally illuminated coleoptile, and between the two zones, it prevented phototropic curvature when it was on the darkened side but had no

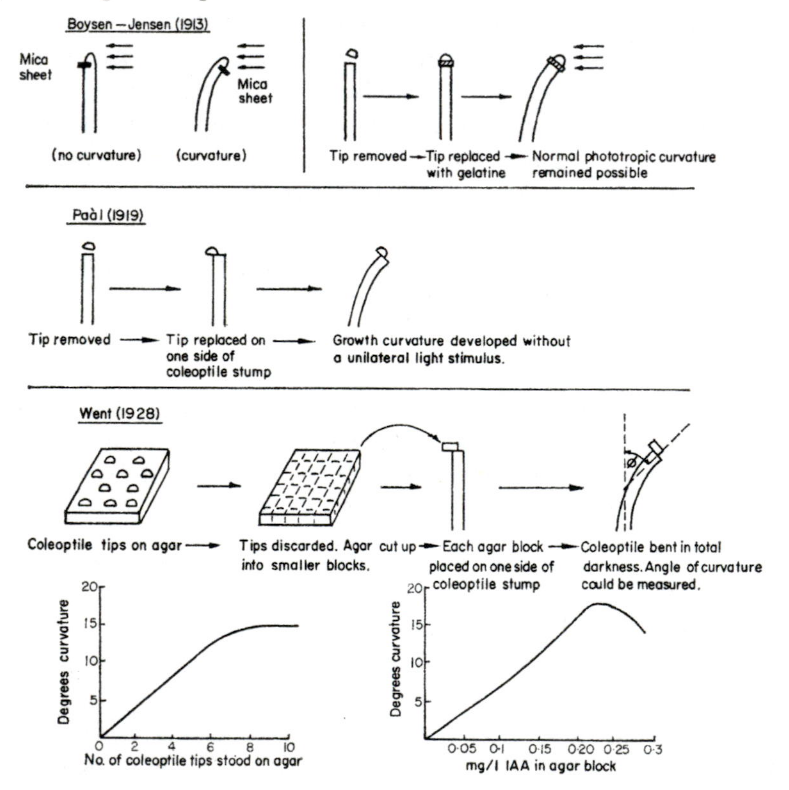

Figure 11.1: Experiments that established the existence of auxin in plants. All used grass coleoptiles. Triple arrows represent the direction of unilateral light. Reproduced by permission of Elsevier, Oxford, from P. F. Wareing and I. D. J. Phillips, *The Control of Growth and Differentiation in Plants* (Oxford: Pergamon Press, 1970).

effect when it was on the illuminated side (Figure 11.1). Cutting did not itself inhibit movement of the chemical because normal phototropism occurred if the coleotile tip was excised and then sealed back onto the stump with a layer of gelatin. Using a slight modification of this last technique, the Hungarian, Paal in 1919 found that the coleoptile tip would induce some curvature in total darkness if it were excised and then replaced asymmetrically on the stump. Growth was promoted on the side of the coleoptile onto which the tip had been placed (Figure 11.1)

Gelatin blocks were the key to the breakthrough made by the young Frits Went, working in the evenings while his days were taken up with national military service. In his own words

> the first experiment applying the diffusate of coleoptile tips into gelatin onto decapitated coleoptiles succeeded at 3am on April 17, 1926, and the next morning (when I had no military duties because it was prince Consort Henry's birthday) I could repeat the experiment to my father's satisfaction.[14]

What Went did was to allow excised coleoptile tips of *Avena* (oat) to rest on gelatin for different lengths of time, cut the gel into smaller blocks and then place the blocks, not the coleoptile tips, onto the coleoptile stumps. He found the block alone was able to initiate the resumption of growth in a darkened tipless coleoptile, but if the block was placed asymmetrically then the coleoptile grew faster on that side (Figure 11.1). The auxin had diffused first from the excised tip into the gel and, then, from the gel into the coleoptile stump. The angle of curvature was proportional to the number of coleoptile tips stood on the gel, so this procedure, using *Avena* coleoptiles, could be used to bioassay, or quantify, the active substance, which was named 'auxin', after the Greek *auxein*, to grow. Barely a year had passed since the death of Francis Darwin when this discovery with such far reaching implications was made.

Went's 1928 thesis became an instant classic, and his *Avena* coleoptile technique became the standard, day-to-day procedure for bioassaying auxin, being sidelined only in the last quarter of the twentieth century when rapid analysis by gas liquid chromatography was available to most laboratories. His thesis unlocked a door to progress. Now it could be demonstrated that when a coleoptile tip that had previously been illuminated from one side is placed on gel blocks separated by a mica sheet, the block on the side that had been shaded contained almost twice as much auxin as the block on the previously lighted side, which is why the former grows more rapidly.

Coleoptiles also provided a clue that would help solve the mystery of gravitropism. When a vertically orientated coleoptile tip was placed on two gel blocks separated by mica there was equal auxin activity in the two blocks. However, if the coleoptile was orientated horizontally, the lower block accumulated twice as much auxin as the upper block. If the same asymmetry occurred in horizontally growing roots, as proved to be the case, this might be the cause of their bending, although the problem remained, why didn't they bend upwards like shoots? The apparent paradox was solved at the California Institute of Technology (Caltech) by Kenneth Thimann,[15] who found concentrations of auxin that *promoted* growth in shoot tissues, such as coleoptiles, *inhibited* growth in root tissues. Thus, in a horizontal root tip, the higher concentration in its lower half would inhibit growth on that side, causing the root to turn downwards.

The link between asymmetry of auxin and the 'Statolith Theory', proposed by Haberlandt and supported by Francis Darwin (see Chapter 8), has only become clear recently. The statoliths, which sediment to the bottom of the cell, induce the local insertion into the plasmamembrane of proteins that pump auxin out of the cell. Whereas, in a vertically growing root auxin flows down the central tissues of the root towards its tip and there is an equal reverse flow away from the tip in the peripheral tissues, in a horizontal root the statoliths divert the reverse flow, causing accumulation in the lower side of the root.[16]

Attempts to extract from either coleotile tips or gelatin blocks (agar later replaced gelatin) enough auxin for the methods of chemical analysis available in the 1920s proved fruitless. The *Avena* test revealed, however, that substances with auxin activity occur widely in nature. One of the richest sources was human urine, and in 1934 it was from urine that auxin was first purified and shown to be indole-3-acetic acid (IAA). IAA was not isolated from a plant source – maize seeds – until the early 1940s.[17]

After completing his PhD., Went worked in the tropics for a brief spell before moving to Caltech where his colleagues included Thimann and number of equally outstanding researchers dedicated to understanding the control of plant growth. At Caltech and other research centres, it became clear that in some situations IAA integrates growth by suppressing rather than stimulating it.[18] Thus, the problem of how plants maintain their overall shape, which had so puzzled Charles Darwin, was explained when it was realized that IAA produced by the apical bud of a shoot will suppress the development of lateral buds on the same shoot, leading to longitudinal extension, rather than lateral expansion, or 'bushing out', of the shoot. So at last it could be understood how the centuries-old practice of pruning, to remove the shoot tips, causes the lower shoots to branch and plants to 'thicken'.

Before long, it became clear that IAA was one, albeit the most important, of a chemical family of naturally occurring substances with broadly similar properties. 'Auxin' then became a generic term for substances with properties like IAA.

And it was not long before the chemists were altering the basic molecule to produce first synthetic auxins, which became known as 'plant growth regulators' because they had, as their name suggests, specific applications in horticulture and agriculture. Among their uses, auxins could be applied to effect floral thinning in over productive orchards and, also, to prevent premature fruit drop during the ripening period. They would promote the formation of roots in cuttings; the naturally occurring indole butryric acid (which is slowly broken down by the plant), and synthetic naphthalene acetic acid (which cannot be broken down by the plant), being especially useful in the vegetative propagation of woody cuttings from shrubs and trees.[19] Synthetic auxins of the phenoxyacetic acid group, such as 2,4, di- and 2,4,5 tri-chlorophenoxyacetic acid (2,4-D and 2,4,5-T, respectively) are amongst the most powerful herbicides known, the latter achieving notoriety during the Vietnam War because it was the major component of 'Agent Orange' sprayed by US aircraft to destroy the native jungle that was providing cover for guerrilla fighters. As more became known about auxins, so other families of plant hormones were discovered. And, more often than not, it was recognized that each process in the growth and maturation of plants is regulated not by one but by two, three, or more hormones – often from different chemical families – acting in concert. Plant hormones, unlike animal hormones, do not have a single 'target' organ or tissue. So, although both auxins and hormones from the gibberellin family are needed for flowering, a high auxin:gibberellin ratio favours the formation of pistillate (male) flowers, while a low ratio favours staminate (female) flowers. Similarly, there is an optimum balance of auxins and cytokinins (hormones regulating cell division) for the successful culture of plant tissues *in vitro*. Such cultures provide an extremely useful way by which relatively slow growing plants of horticultural importance, such as fruit bushes, can be propagated rapidly and in large numbers. *In vitro* cultures are employed also in the propagation of many non-woody crops, such as potato, where the objective is to produce disease-free stock by culturing healthy tissues excised from infected but otherwise highly valued parental material.

Commercial interest in plant growth regulators fired interest in, and no doubt increased the funding for, further pure research. One of the most significant outcomes of such pure research was the gradual accumulation of evidence concerning that most fundamental of processes, the expansion of young cells. The final shape of the mature plant cell depends on its wall which is, effectively, a rigid external skeleton preventing any further turgor-driven expansion. It is now recognized that auxin promotes the transverse orientation of cellulose microfibrils that are laid down within the cell wall during its extension, while ethylene, a gaseous plant hormone, promotes the longitudinal orientation of fibrils (Figure 11.2). Thus, a high auxin:ethylene ratio leads to long thin cells, while a low ratio leads to short fat cells. Just to confuse matters, exogenous application of auxin – which adds to and, thereby, raises endogenous auxin to abnormally high

(A) Randomly oriented cellulose microfibrils

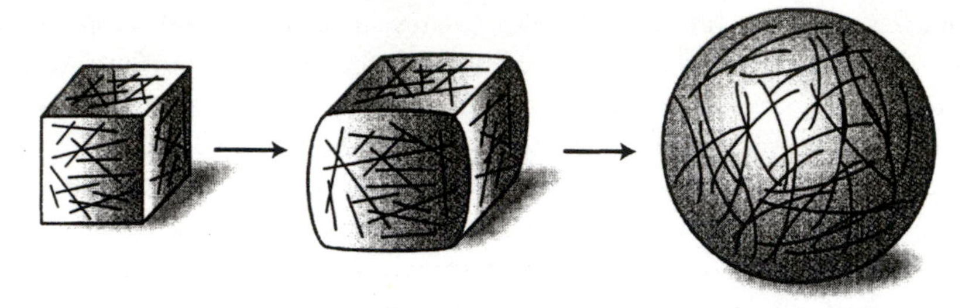

(B) Transverse cellulose microfibrils

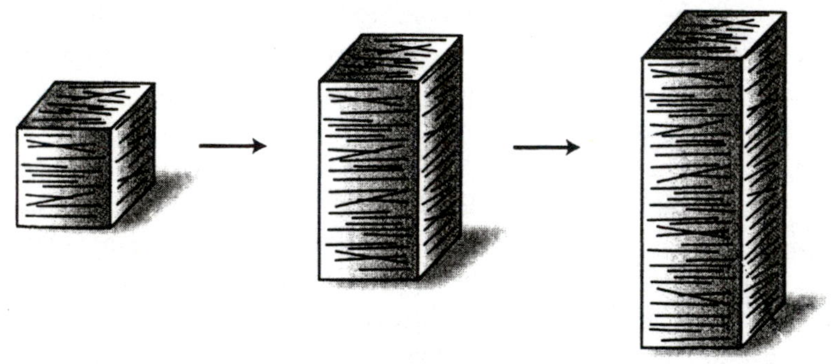

Figure 11.2: The orientation of newly deposited cellulose microfibrils determines the direction of cell expansion. (A) The cell wall is reinforced by randomly orientated microfibrils and assumes a spherical shape. (B) Most microfibrils have a transverse orientation and the cell expands longitudinally, lateral expansion being constrained. Reproduced by permission of Sinauer Associates, US, from L. Taiz and E. Zeiger, *Plant Physiology*, 4th edn (Sunderland, MA: Sinauer Inc., 2006).

concentrations – can stimulate tissues to produce ethylene, with the result that affected cells are short and fat, and tissues are swollen.

*****

The practical importance of what followed from Francis's groundbreaking studies of stomatal movements and his encouragement of the young F. F. Blackman (see Chapter 8) was enormous. It led to the birth of crop physiology, a whole new area of research which – with the collaboration of plant breeders –was dedicated to improving the photosynthetic efficiency of crops while at the same time improving their drought resistance. If that were not significant enough, it is now clear that the Francis Darwin-F. F. Blackman thread is directly connected to our twenty-first century efforts to understand interactions between plants and climate. Whether the subject is crops or wild plants, attention is focussed on photosynthesis and transpiration.

What links transpiration and photosynthesis is diffusion – the movement of molecules by random thermal agitation. It was Adolf Fick, Professor of (Animal) Physiology and colleague of Sachs at Würtzburg, who in the 1880s provided the basis for our modern understanding. Trained as a physicist, Fick studied the general phenomenon of diffusion, arriving at a series of physical laws with application to biology. His first law states that the rate of diffusion of a substance, such as CO_2 or water vapour, between two points, say the inside and the outside of a stomatal pore, is directly proportional to the difference in its concentration between the two points.[20] The rate of transport of the substance, as through a pore, depends upon temperature, the cross sectional area of the pore, and the diffusion coefficient of the substance. Larger molecules have lower diffusion coefficients than smaller molecules (where low=slow); thus CO_2 with a molecular weight of 44 has a lower diffusion coefficient than water, with a molecular weight of 18, and diffuses more slowly than water.

The first, but transient, biochemical product of carbon fixation in photosynthesis is for most plants the three-carbon compound, 3-phosphoglyceric acid. These 'C3 plants' lose about 500 molecules of water for every one molecule of carbon dioxide that they fix. Rather more efficient in their use of water, by virtue of the modified internal anatomy of their leaves, are the 'C4 plants', mainly grasses such as maize, whose first product of photosynthesis is the four-carbon compound oxaloacetic acid. Still more efficient are plants exhibiting Crassulacean Acid Metabolism – CAM plants – for these typical desert dwellers open their stomata at night when evaporative stress is minimal. They fix carbon in a specialized two-stage process.

In C3 plants the large ratio of water efflux to CO_2 influx is explained by Fick's Law in two ways:

- CO_2 molecules diffuse about 1.6 times more slowly through air than water molecules do
- water molecules are highly concentrated in the air spaces within the leaf, while the concentration of CO_2 in ambient air is still only 0.0382 per cent (382 pp), so the concentration gradient driving water loss is about 50 times larger than that driving CO_2 uptake.

Diffusion coefficients for substances in a gaseous phase, such as air, are much greater than diffusion coefficients for the same substances in water, which becomes significant for CO_2 molecules entering leaves because they have one more step to negotiate than do water molecules leaving the leaf (Figure 11.3). In their pathway from the atmosphere to the point of chemical fixation within leaf cells, CO_2 molecules have to diffuse through the aqueous solution of the cell and its water-saturated wall. The effect of this 'mesophyll resistance' is to raise the concentration of CO_2 inside the leaf, so reducing the relatively small concentration gradient along the stomatal diffusion pathway – though in C4 plants the effect is mitigated somewhat by their peculiar anatomy.

The ratio between carbon gain and water loss can been expressed in many ways, one of the most popular being 'water use efficiency' (WUE), a parameter that can be variously calculated for anything from a leaf to a forest. What emerges is that, in the same given environment, WUE is remarkably constant for a particular plant type, C3, C4 or CAM, across a wide range of species or cultivars. Numerous studies confirm the close link between atmospheric humidity and transpiration rate, as is predicted by Fick's Law. And numerous studies point to the importance of stomatal behaviour in regulating water loss. For example, those species showing partial closure around mid-day, when ambient air is typically at its driest, tend to have high WUEs.

In C3 and C4 plants, with an adequate supply of water, stomata open during the first hours of daylight and close, more slowly, during the last hours of the day (earlier if water is in short supply), just as Francis Darwin and Dorothea Pertz first found. At all but the widest apertures – occurring normally in well watered plants in a high light environment – where photosynthesis is limited by the mesophyll resistance, transpiration and assimilation are linearly related to stomatal aperture.[21] This is the theory which invited the possibility that transpiration in field crops might be reduced, at least through short periods of drought, by the application of antitranspirants, synthetic chemicals that physically block the

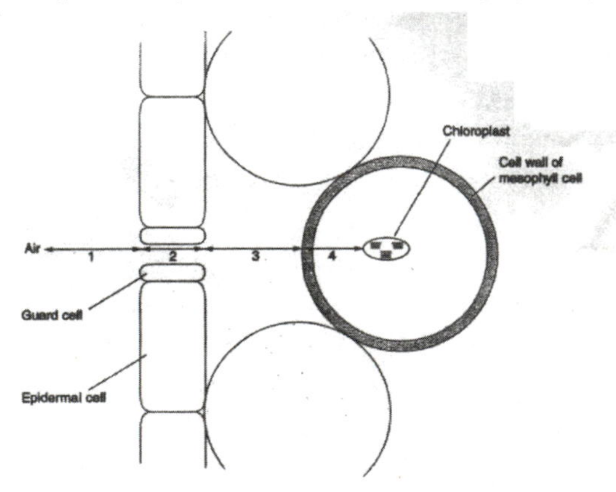

Figure 11.3: Diagrammatic section through a leaf surface (not to scale) to show the diffusion pathways. Air spaces, occupying 14–40 per cent of the internal volume of a leaf, are saturated with water vapour evaporating from the wet surfaces of the cell walls. CO_2 entering the leaf diffuses from ambient air to the site of fixation within the leaf (the chloroplast) encountering four resistances on its way: 1 = boundary layer of unstirred air; 2 = stomatal resistance; 3= internal resistance; 4= mesophyll resistance (diffusion through water). Water vapour leaving the leaf encounters only 1, 2 and 3, so diffusion is solely through air. It was one of Blackman's collaborators, E. J. Maskell, who first suggested the pathways could be represented as a number of resistances in series.

stomatal aperture, or abscisic acid (ABA), a naturally occurring plant hormone that induces stomatal closure. What has been found in practice is that any slowing of transpiration tends to cause the leaf to get hotter – evapotranspiration having a cooling effect – which, in turn, increases the concentration gradient of water across the stomatal pores, having the undesired effect of accelerating transpiration. In spite of receiving much attention, ABA has proved less effective in controlling transpiration from field crops than was hoped.[22]

Another practical application to emerge from this field of study, and an indubitably successful one, originated with E. J. Maskell working in F. F. Blackman's Cambridge laboratory in 1928. While measuring CO_2 assimilation in leaves of cherry laurel under constant illumination, Maskell[23] simultaneously measured stomatal apertures. He was able to confirm that at near-normal concentrations of CO_2 (about 300 ppm) the pattern of assimilation was closely linked to the rhythm of stomatal opening that persisted in continuous light. However, this linkage was not apparent when CO_2 was supplied at concentrations as high as 2.3 per cent: for, even when stomata were relatively closed, enough CO_2 could diffuse into the leaf to support a high rate of assimilation. When it was seen that increasing CO_2 concentrations above normal ambient levels resulted in increased assimilation, in spite of the fact that higher concentrations induce smaller stomatal apertures, it was realized that crops might benefit from 'CO_2 feeding'. Such feeding has proved extremely effective both biologically and economically in closed environments such as glasshouses and today it is standard horticultural practice to enhance CO_2 concentrations around glasshouse crops such as lettuce, tomatoes and cucumbers in order to enhance their growth. The gas may be released directly from storage cylinders, or, less expensively, generated by the combustion of a hydrocarbon fuel.

Many of those early authors who wrote about the physiology of plants devoted lengthy sections of their books to the reactions of plants to strong acids and alkalis, and to the effects of toxic chemicals, such as mercury and other heavy metals. Their interest lay both in listing environmental threats for their own sake and in relating the plant's responses to the much wider subject of 'irritability'. The latter responses included plant movements, such as those of the sensitive plant, *Mimosa*, in response to touch. It is appropriate therefore that Francis Darwin should provide the key that has allowed modern botanists to identify a firm link between movements (of stomata) and environmental toxins (in the form of air pollutants). The connection exists because carbon dioxide is, of course, not unique in being able to diffuse through stomatal pores into the air spaces with the leaf. Pollutant gases, such as sulphur dioxide and the oxides of nitrogen that are part of vehicle exhaust fumes – and minor combustion products of glasshouse stoves – unfortunately, also get into the leaf via stomata.[24] Improved knowledge of the critical concentrations for each gas, singly or as mixed cocktails, has resulted very largely from an improved understanding of their diffusion through stomatal pores. Although crop plants can be bred for increased resistance, natural vegetation cannot and the main outcome of over a quarter of a century of research had been better informed legislation to restrict the anthropogenic release of pollutant gases.

Physiological measurements of assimilation and transpiration have not only became an essential part of agronomic studies of the effects of irrigation or nutrition – informing breeding programmes and often defining their goals – but they have become central to arguments about the causes and potential solutions to problems of climate change. Models of global photosynthesis strongly suggest that the distribution of carbon fixation parallels the distribution of transpiration – with highest fluxes of both carbon and water occurring in vegetation in the wet tropics – indicating that stomatal control of both processes is tightly coupled.[25] Remembering that global agriculture now accounts for 70 per cent of the water used by humans,[26] and that in developed countries crop losses due to lack of water exceed losses due to all other causes combined, it is clear that predicted changes in rainfall patterns will have major consequences for the capacity of plants to buffer CO_2 levels in the world's atmosphere.

Finally, and appropriately for a book about the Darwins, evidence is rapidly accumulating that there has been an *evolutionary* interaction between stomata and the concentration of CO_2 in the atmosphere. Examination of dried herbarium specimens collected during the industrial age (the last 200 years) shows that as CO_2 concentrations have risen stomatal frequencies have declined, by as much as 40 per cent in some tree species.[27] On the much longer, geological, time scale evolution of the differently shaped stomates of grasses, with their dumb-bell shaped guard cells, had a profound effect on the history of the earth's vegetation. Such stomata have improved sensitivity to small changes in turgor, enabling grasses to react much more swiftly than the majority of plants, with kidney shaped guard cells (see figure 1.1), to changes in water supply. This was a major factor contributing to the relatively high drought tolerance of grasses. However, of even greater importance on this longer time scale was the evolution of the C4 grasses, which combined both drought tolerance and high photosynthetic efficiency. Beerling[28] has argued that C4 plants evolved at a time when concentrations of CO_2 in the atmosphere were plummeting and C3 plants were needing to keep their stomata open for longer and longer to capture CO_2, with the risk of losing more and more water. What is beyond question is that this small group has been highly successful: C4 grasses cover more than 20 per cent of vegetated land surface, account for 30 per cent of terrestrial vegetated land surface, and include maize and sugarcane, two of the most productive crops known. The down side is that C4 grasslands have eliminated large areas of forest, in part because although dried, dead grassland burns readily, it also regenerates readily – much more quickly than trees. The smoke and particulates from fires serve to depress local rainfall further, hastening the decline of the forests and impacting on wider global patterns of rainfall.

A century after Francis's pioneering studies of stomatal movements, using the most primitive porometers, those who came after him have been led in directions far beyond his experience and towards complexities that they are far from resolving.

The practical benefits of plant physiology, as with animal physiology, took longer to be delivered than the pioneers hoped but by the mid-twentieth century they were clear to see. In at least two areas of the utmost practical importance the legacy of Charles and Francis proved invaluable.

*****

Where did the *personal* trail lead from Francis, and can he be said to have founded a Cambridge School of Plant Physiology analogous to Michael Foster's School of (Animal) Physiology? In the absence of the formal leadership and administrative structures that characterize a body such as The Botany School, the first criterion for a less formal grouping to be called a 'School' is that it should have complementary, if not collaborating, researchers. Secondly, through postgraduate teaching and experience, it should train sufficient graduates to maintain and, preferably, to grow the School – even to the point where its methods and ethos are exported to other institutions. Additionally, but not necessarily, it should teach to undergraduates those discipline(s) which are the School's defining strengths.

It is Sydney Vines, rather than Francis, who has the best claim to be the founder of a School, if such existed, for it is he who started plant physiology teaching and research in Cambridge (Chapter 8). However, these activities would almost certainly have withered if Francis had not first assisted Vines and then developed his own courses. Co-incident with Francis there were his colleagues Gardiner and Marshall Ward, each pursuing his own physiological interests. Soon, under Francis's guidance, F. F. Blackman was added.

There were collaborations with Bateson, the geneticist, and with Rowland Biffen, the plant breeder, each destined to have huge impacts on British crop science. While the former was the first Director of the John Innes Institute, then at Merton in Surrey, the latter was appointed a Demonstrator in the Botany School in 1898 before joining the newly founded Cambridge School of Agriculture in 1899. Biffen was among the first to perceive the opportunities offered to plant breeders by the new understanding of heredity; in 1912 he was made Director of the newly established Plant Breeding Institute near Cambridge. (In moving the emphasis from pure to applied research, Britain was soon moving in the opposite direction to the United States. Between 1912 and 1926, and driven by food shortages during World War I, the British government founded research institutes in the applied disciplines of plant breeding, animal nutrition, animal pathology, horticulture and plant viruses).[29] The two institutes, started by Bateson and Biffen, were finally merged in 1994, at Norwich, forming the world famous John Innes Centre.

In relation to the second criterion of a School, Francis's record in training postgraduates who went on to become distinguished physiologists was – like that of his contemporaries, Vines, Ward and Gardiner – indifferent. In particular, none of Francis's several female assistants was to dedicate her life to a career in plant physiology (see Chapter 9). Francis can, however, take credit for teaching plant physiology to F. F. Blackman, and to his younger brother Vernon who became Professor of Vegetable Physiology at Imperial College, London. It was through 'F. F.' that Francis's person-to-person connections and the Botany School's involvement in plant physiological research were maintained. Appointed to a Readership when Francis resigned his post in 1904, Blackman's own list of research students included T. A Bennet-Clark, F. Kidd and G. E. Briggs. The first two made major contributions to our understanding of the actions of plant hormones (auxin and ethylene, respectively), while Briggs pioneered studies of solute movements across the membranes and walls of plant cells. The latter, with Kidd and West, published in 1920 a seminal paper that launched plant 'growth analysis'.[30] Based on Vernon Blackman's idea, it demonstrated that plants grow exponentially, rather like capital invested for compound interest in a bank account. With applications to agronomy and plant breeding, growth analysis enabled the photosynthetic performance of either different management practices, or of new cultivars, to be quantified so enabling more precise comparisons. F. F. Blackman's interest in respiration was carried forward by one of his students, W. O. James, who took a more biochemical approach, studying the metabolic pathways and enzymes of respiration. James was to become a leading figure in the new world of plant biochemistry, particularly after his move to Oxford.

Cambridge taught plant physiology to undergraduates before most other universities in Britain recognized the discipline. Consonant with the philosophy of the New Botany there was a strong emphasis on the practical component of teaching. Darwin and Acton's textbook, which was refined from the experiences of those classes, became the national, or even international, standard for laboratory classes for more than a generation (see Chapter 8).

A sub-department of plant physiology was recognized in Cambridge in 1937, but was there an effective 'School' of plant physiology in Francis's lifetime? The numbers of staff never matched those in Foster's group, but this was to be expected because botany, having divorced itself from medicine, was and would remain a much smaller discipline. The numbers of postgraduates and, particularly, of visiting scholars working on a sabbatical-like basis never matched the numbers passing through the laboratories of Sachs[31] or Pfeffer,[32] so person-to-person connections were much fewer. Nevertheless, a strong plant physiological tradition was undoubtedly established. There is on balance a reasonable claim therefore that, through teaching and the sheer quality of the research that emerged from Cambridge before and immediately after Francis's death, he was part of a small but influential School.

*****

The influence of F. F. Blackman in one other direction is of historical interest. Although physiology was firmly established in the undergraduate teaching curriculum for advanced or honours students in Cambridge thanks to Vines and Francis Darwin, it was, thought Blackman and his sympathizers (or co-conspirators), insufficiently represented in the elementary course. The position was, they believed, no better in other British Universities. Some evidence for this is gained by looking at the Elementary Course that Marshall Ward taught in 1904, the year after Francis left the Botany School. In the Michaelmas Term there was organography, morphology and anatomy, and physiology; in the Lent Term, an evolutionary course on the biology of fungi, algae, bryophyta, vascular cryptogams and conifers; and in the Easter Term, the systematics of flowering plants. Ward's successor, A. C. Seward, a palaeobotanist, changed things little.

Frustrations multiplied and in 1917, F. F. Blackman, together with his brother Vernon, brother-in-law, Arthur Tansley, and two other leading botanists published in *New Phytologist* a controversial letter, 'The Reconstruction of Elementary Botanical Teaching', which came to be known as the 'Tansley Manifesto'. Tansley's name was associated with the letter largely because he was the founder and Editor of *New Phytologist* – and he may not have been unhappy at the attention the article drew to his journal – although recent scholarship points to the Blackmans as the driving force behind the letter.[33] And draw attention, it certainly did. Tansley invited responses and published a selection of the many he received.

The Manifesto argued

> Botany in this country is still largely dominated by the morphological tradition, founded on an attempt to trace phylogenetic relationships of plants, which began as the result of the general acceptance of the doctrine of descent. Elementary teaching (as well as a very large part of advanced teaching) is mainly occupied with the endless facts of structure and their interpretation from the phylogenetic standpoint. Side by side with this there generally goes a discussion which is often limited by a crude Darwinian teleology. Plant physiology is relegated in most cases to a subordinate place and is taught as a separate subject. The newer studies of ecology and genetics play a very small part in the curriculum.[34]

Pouring petrol on their fire, they opined, the current morphological approach did not 'attract the best types of mind amongst possible students'.[35] Nothing could have been calculated to anger the 'morphologists' more.

There was as much pedantry as there was substance in their responses. Led by F. O. Bower (Glasgow) and directed from behind the scenes by Isaac Bayley Balfour (Edinburgh), the morphologists' central argument was that the widely taught organography had been initiated by Sachs himself and that it combined a comparative treatment of plant parts with a description of their functioning. The letters and arguments pro- and contra- the Manifesto were analysed in detail by Don Boney who concluded that 'the dedicated revolutionaries of the 1870s and early 1880s [such as Bower], the ardent protagonists of the 'New Botany', became in later life ... entrenched and inflexible ... confirmed conservatives'. Boney found that not only was the balance of letters received by Tansley broadly in favour of the Manifesto but, more importantly, its supporters came 'from individuals newly appointed to chairs, or who were soon to be professors, many of whom in the next 20–30 years would make significant impacts on those branches of botanical science, physiology and ecology, then under discussion'.[36] The 'Tansley Manifesto' – should it have been the Blackmans' Manifesto? – challenged botanists to reflect on what should be at the core of botany teaching. For several decades afterwards the answer proved to be plant physiology. The groundwork of Vines and Francis Darwin bore fruit.

*****

To leave the history of botany at this juncture is justifiable on grounds other than Francis's death, for in the late 1920s not only were the new disciplines of ecology and biochemistry energetically asserting their independence from physiology but the world of botany was reaching another watershed, one determined by logistics.

Europe was impoverished by the loss of so many of its best young men. The Darwin families were exceptional in being spared the grief of losing a son on the battlefields of World War I. Many were not so lucky.[37]

Britain's brief period of domination in botanical science was ending. Simply in terms of numbers of funded researchers – who by their very existence generated a virtuous circle of more new ideas and projects – leadership was inexorably passing from east to west across the Atlantic. The nineteenth-century German model of a university system with authoritarian professors, selectivity and centralization had not been immediately transferable to the more democratic, non-selective institutions of North America.[38] Excepting a few universities, like Harvard, where pure botany had always been valued for its own sake, colleges in most states were, in their early years, orientated towards applied botany, tackling practical problems that affected local farmers. Asa Gray and, particularly, Will Farlow may have imported the New Botany directly from Germany to Harvard but they were the exceptions, for the most part New Botany reached North America indirectly, being modified on its way by the British experience. However, by the mid 1920s, America was ready to take on the mantle of leadership in pure as well as applied botany. The American Society for Plant Physiology was formed in 1925 and a year later published the first volume of its journal, *Plant Physiology*, soon to prove a world leader.

*****

A century has passed since the last practical contribution of the Darwin line to botany. What they thought and did would have amounted to little if it had not been picked up and taken further by others, such as Blackman and Went. But it was, and the Darwins' contribution to the development of the science of botany has been immense.

It is salutary to reflect that today's plant scientists may have the capacity to analyse the sequence of chemical bases in the DNA of any plant, and to map all the genes that the plant possesses, but as they now enter the post-genomic age they face many problems that would have been immediately recognized by the Darwins. In 2007, an editorial in *New Phytologist*, which has been publishing the best botanical research papers for 105 years, noted, 'Today the skills of plant physiologists and developmental biologists are at a real premium and are urgently required to unlock the secrets of the treasures uncovered during the postgenomic era'.[39] The focus of botanists has shifted back from the technicalities of analysing genes and DNA to problems, such as floral development, root hair deformation, root growth and drought tolerance that the Darwins would have recognized. Their botany remains at the core of the subject.

*****

> Book material is of a gaseous or theoretical nature, unlike plant material which is solid fact.[40]

NOTES

1 Green Threads across the Ages

1. D. Beerling, *The Emerald Planet: How Plants Changed Earth's History* (Oxford: Oxford University Press, 2007).
2. C. R. Darwin, *The Life and Letters of Charles Darwin*, ed. F. Darwin, 3 vols (London: John Murray, 1887), vol. 1, p. 98; vol. 2, p. 99.
3. D. King-Hele, *Erasmus Darwin: A Life of Unequalled Achievement* (London: De La Mare, 1998).
4. C. Körner, 'Plant CO_2 Responses: An Issue of Definition, Time and Resource Supply', *New Phytologist*, 172 (2006), pp. 393–411.
5. E. A. Ainsworth and S. P. Long, 'What Have We Learned from 15 years of Free-Air CO_2 Enrichment (FACE). A Meta-Analytic Review of the Responses of Photosynthesis, Canopy Properties and Plant Production to Rising CO_2', *New Phytologist*, 165 (2005), pp. 351–72.
6. J. von Sachs, *History of Botany (1530–1860)*, trans. from the German edition by H. E Garnsey and I. B Balfour with an original preface (Oxford: Clarendon Press, [1875] 1906), p. xii.

2 The Fortunes of the Darwins

1. J. Browne, *Charles Darwin: Voyaging. Volume I of a Biography* (London: Pimlico, 1995); J. Uglow, *The Lunar Men: The Friends Who Made the Future 1730–1810* (London: Faber and Faber, 2002).
2. D. King-Hele, *Erasmus Darwin* (London: Macmillan, 1963); King-Hele, *Life of Unequalled Achievement*.
3. King-Hele, *Life of Unequalled Achievement*.
4. Uglow, *Lunar Men*
5. King-Hele, *Erasmus Darwin*, p. 14.
6. King-Hele, *Life of Unequalled Achievement*.
7. E. Krause, *Erasmus Darwin ... With a Preliminary Notice by Charles Darwin* (1879), excerpted in The *Works of Charles Darwin*, ed. P. H. Barrett and R. B. Freeman, 29 vols (London: Pickering & Chatto, 1988–9), vol. 29.
8. Browne, *Charles Darwin: Voyaging*.
9. J. Browne, *Charles Darwin: The Power of Place. Volume II of a Biography*. (London: Jonathan Cape, 2002).

10. King-Hele, *Erasmus Darwin*; King-Hele, *Life of Unequalled Achievement*.
11. Browne, *Charles Darwin: The Power of Place*.
12. C. R. Darwin, *On the Origin of Species by Means of Natural Selection, or the Preservation of Favoured Races in the Struggle for Life* (1859), reprinted in *Works of Charles Darwin*, vol. 15.
13. A. Tennyson, 'In Memoriam' (1850), canto 56.
14. C. R. Darwin, *The Origin of Species by Means of Natural Selection, or the Preservation of Favoured Races in the Struggle for Life*, 6th edn (1876), reprinted in *Works of Charles Darwin*, vol. 16, pp. 168–9.
15. T. Veak, 'Exploring Darwin's Correspondence; Some Important but Lesser Known Correspondents and Projects', *Archives of Natural History*, 30 (2003), pp. 118–38.
16. B. Darwin, *The World that Fred Made* (London: Chatto and Windus, 1955); E. Healey, *Emma Darwin: The Inspirational Wife of a Genius* (London: Hodder Headline, 2001).
17. B. Darwin, *The World that Fred Made*.
18. Browne, *Charles Darwin: The Power of Place*, p. 235.
19. B. Darwin, *The World that Fred Made*, p. 84.
20. Ibid., pp. 53–4.
21. Ibid., p. 54.
22. Healey, *Emma Darwin*, p. 297.
23. Browne, *Charles Darwin: The Power of Place*, pp. 446, 436, 437.
24. Healey, *Emma Darwin*; R. B. Freeman, 'The Darwin Family', *Biological Journal of the Linnean Society*, 17 (1982), pp. 9–21.
25. F. Darwin, 'Reminiscences of my Father's Everyday Life' (1884), in C. R. Darwin, *The Autobiography of Charles Darwin 1809–82* (New York: Barnes and Noble Books, 1958), p. 97; B. Darwin, *The World that Fred Made*, p. 27.
26. *Life and Letters of Charles Darwin*, vol. 3, p. 309
27. Ibid., vol. 3, p. 332.
28. E. Bunning, *Ahead of his Time: Wilhelm Pfeffer. Early Advances in Plant Biology*, trans. H. W. Pfeffer (Ottawa, ON: Carleton University Press, 1989).
29. Uglow, *Lunar Men*.
30. P. O'Brian, *Joseph Banks: A Life* (London: Harvill Press, 1987).
31. King-Hele, *Life of Unequalled Achievement*.
32. C. R. Darwin, *Autobiography*, p. 84.
33. Browne, *Charles Darwin: The Power of Place*.
34. *Life and Letters of Charles Darwin*, vol. 1, p. 22.
35. Ibid., vol. 1, p. 36.
36. Browne, *Charles Darwin: Voyaging*, p. 374.
37. F. Darwin, 'Recollections', in *Springtime and Other Essays* (London: John Murray, 1920), pp. 51–71; p. 55.
38. Ibid., p. 54.
39. F. Spalding, *Gwen Raverat: Friends, Family and Affections* (London: Harvill Press, 2001), p. 163.
40. G. Raverat, *Period Piece: A Cambridge Childhood* (London: Faber and Faber, 1954), p. 147.
41. A. Briggs, 'The Later Victorian Age', in B. Ford (ed.), *The Cambridge Cultural History of Britain* (Cambridge: Cambridge University Press, 1995), pp. 2–38.

3 The Misfortunes of Botany

1. J.-J. Rousseau, *Letters on the Elements of Botany Addressed to a Lady (1771–3)*, trans. T. Martyn (London: B White, 1785), p. 1.
2. A. B. Shteir, *Cultivating Women, Cultivating Science: Flora's Daughters and Botany in England 1760– 1860* (Baltimore, MD: Johns Hopkins University Press, 1996).
3. J. E. Smith, *Introduction to Physiological and Systematical Botany*, 3rd edn (London: Longman, 1813), p. xvii.
4. E. Lankester, Review of *An Elementary Course in Botany* by A. Henfrey, *Athenaeum* (3 October 1857), p. 1562.
5. J. M. Mackenzie, 'Plant Collecting and Imperialism', in J. Illingworth and J. Routh (eds.), *Reginald Farrar: Dalesman, Planthunter, Gardener* (Lancaster: Centre for North-West Regional Studies, University of Lancaster, 1991), pp. 8–14; C. Lyte, *The Plant Hunters* (London: Orbis, 1983).
6. King-Hele, *Life of Unequalled Achievement*; Uglow, *Lunar Men*.
7. King-Hele, *Erasmus Darwin*.
8. King-Hele, *Life of Unequalled Achievement*.
9. Shteir, *Cultivating Women*.
10. D. King-Hele, 'The 1997 Wilkins Lecture: Erasmus Darwin, the Lunaticks and Evolution', *Notes and Records of the Royal Society of London*, 52 (1998), pp. 153–80.
11. E. Darwin, *The Loves of the Plants; A Poem with Philosophical Notes* (London: Jones & Co., 1825), canto IV, ll. 335–44.
12. Shteir, *Cultivating Women*.
13. Uglow, *Lunar Men*.
14. King-Hele, *Life of Unequalled Achievement*.
15. E. Darwin, *Loves of the Plants*, p. 135.
16. A. H. Dupree, *Asa Gray 1810–1888* (Cambridge, MA: Harvard University Press, 1959), p. 28.
17. Von Sachs, *History of Botany*, p. 89.
18. J. A. Secord, 'Artisan Botany', in N. Jardine, J. A. Secord and E. C. Spray (eds), *Cultures of Natural History* (Cambridge: Cambridge University Press, 1996), pp. 378–93.
19. W. J. Hooker, *The British Flora: Vol. I. Comprising the Phænogamous or Flowering Plants, and the Ferns* (London: Longman, 1842).
20. King-Hele, 'Erasmus Darwin, the Lunaticks and Evolution'.
21. E. Darwin, *Phytologia or the Philosophy of Agriculture and Gardening with the Theory of Draining Morasses and with an Improved Construction of the Drillplough* (London: J. Johnson, 1800).
22. E. Darwin, *The Botanic Garden. A Poem in Two Parts; Containing The Economy of Vegetation and the Loves of The Plants* (London: Jones & Co., 1825), p. vii.

4 Erasmus Darwin's Vision of the Future: *Phytologia*

1. E. J. Russell, *A History of Agricultural Science in Great Britain 1620–1954* (London: George Allen and Unwin, 1966).
2. M. J. Schleiden, 'Beiträge zur Phytogenesis', *Archiv für Anatomie, Physiologie, und Wissenschaftiche Medicin*, 13 (1838), pp. 137–76.

3. T. Schwann, *Mikroskopische Untersuchungen über die Übereinstimmung in der Struktur und dem Wachstum der Tiere und Pflanzen* (Berlin: Sander'schen Buchhandlung, 1839).
4. J. Farley, *Gametes and Spores* (Baltimore, MD: Johns Hopkins University Press, 1982).
5. A. G. Morton, *History of Botanical Science* (London: Academic Press, 1981).
6. Uglow, *Lunar Men.*
7. E. Darwin, *Phytologia*, p. vii.
8. Ibid., p. 1.
9. The Reverend Stephen Hales was Perpetual Curate of the Parish of Teddington, near London, and Rector of Farringdon, Hampshire (where he was a neighbour of Gilbert White of Selbourne). Among his many interests, Hales campaigned against the distillation and consumption of gin, designed ventilators to improve the atmosphere in ships and prisons, was a Trustee of the Colony of Georgia, and helped to found the Royal Society of Arts. See A. E. Clark-Kennedy, *Stephen Hales, D.D., F.R.S.* (Cambridge: Cambridge University Press, 1929); I. B. Cohen, 'Stephen Hales', *Scientific American*, 234 (1976), pp. 98–107.
10. Clark-Kennedy, *Stephen Hales*; Cohen, 'Stephen Hales'.
11. S. Hales, *Vegetable Staticks*, reprinted from the 1st edn with a foreword by M. A. Hoskin (London: The Scientific Book Guild, [1727] 1961), p. xxxi.
12. The verses are by Thomas Twining, a contemporary; cited in Hales, *Vegetable Staticks*.
13. Clark-Kennedy, *Stephen Hales*; Cohen, 'Stephen Hales'.
14. D. C. G. Allan and R. E. Schofield, *Stephen Hales: Scientist and Philanthropist* (London: Scholar Press, 1980).
15. The Reverend Joseph Priestley was a Dissenting minister. He had a lopsided face, walked with a bird-like trot and talked non-stop. After his best scientific work was done, he was forced to flee to Pennsylvania because his opponents thought he was trying to import the French Revolution to Britain. See Uglow, *Lunar Men*; F. W. Gibbs, *Joseph Priestley: Adventurer in Science and Champion of Truth* (London: Nelson, 1965).
16. Ibid.
17. Allan and Schofield, *Stephen Hales*.
18. Uglow, *Lunar Men*, p. 237.
19. Gibbs, *Joseph Priestley*; A. D. Krikorian, 'Excerpts from the History of Plant Physiology and Development', in P. J. Davies (ed.), *Historical and Current Aspects of Plant Physiology: A Symposium Honoring F. C. Steward* (New York: New York State University, 1975), pp. 9–97.
20. King-Hele, *Life of Unequalled Achievement*.
21. Uglow, *Lunar Men*.
22. N. Beale and E. Beale, *Who was Ingen Housz, Anyway?: A Lost Genius* (Calne: Calne Town Council, 1999).
23. Ingen-Housz found that in a growing plant daily CO_2 uptake by green tissues in the light was greater than the amount of CO_2 released by green and non-green tissues in light and darkness. H. S. Reed, 'Jan Ingenhousz Plant Physiologist. With a History of the Discovery of Photosynthesis', *Chronica Botanica*, 11 (1949), pp. 285–396; P. Smit, 'Jan Ingen-Housz (1730–1799): Some New Evidence about his Life and Work', *Janus*, 117 (1980), pp. 125–39.
24. Hales, *Vegetable Staticks*, p. 186.

25. Beale and Beale, *Who was Ingen Housz, Anyway?*; H. Gest, 'Bicentenary Homage to Dr Jan Ingen-Housz, MD (1730–1799), Pioneer of Photosynthesis Research', *Photosynthesis Research*, 63 (2000), pp. 183–90.
26. Smit, 'Jan Ingen-Housz'.
27. *Life and Letters of Charles Darwin*.
28. Gibbs, *Joseph Priestley*; Reed, 'Jan Ingenhousz'.
29. J. Sénebier, *Physiologie Végétale* (Geneva: J. J. Paschoud, 1782).
30. Von Sachs, *History of Botany*.
31. H. Gest, 'A "Misplaced Chapter" in the *History of Photosynthesis Research*; the Second Publication (1796) on Plant Processes by Dr Jan Ingen-Housz, MD, Discoverer of Photosynthesis', *Photosynthesis Research*, 53 (1997), pp. 65–72.
32. E. Darwin, *Phytologia*, p. 45.
33. Ibid., p. 77.
34. H. Hart, 'Nicolas Theodore de Saussure', *Plant Physiology*, 5 (1930), pp. 424–9.
35. Morton, *History of Botanical Science*; von Sachs, *History of Botany*.
36. F. W. J. McCosh, *Boussingault: Chemist and Agriculturalist* (Dordrecht: Kluwer, 1984).
37. E. Darwin, *Phytologia*, p. 244.
38. Ibid., p. 195.
39. Sir H. Davy, *Elements of Agricultural Chemistry* (London: W. Bulmer and Co., 1813), p. 310.
40. Russell, History of Agricultural Science; R. C Burns and R. W. F Hardy, *Nitrogen Fixation in Bacteria and Higher Plants* (Berlin: Springer-Verlag, 1975).
41. Von Sachs, *History of Botany*.
42. Krikorian, 'Excerpts from the History of Plant Physiology and Development'; A. D. Hall, *The Book of Rothamsted Experiments* (London: John Murray, 1905).
43. McCosh, *Boussingault*.
44. J. B. Lawes, J. H. Gilbert and E. Pugh, 'On the Sources of the Nitrogen of Vegetation; with Special Reference to the Question Whether Plants Assimilate Free or Uncombined Nitrogen' in *Philosophical Transactions of the Royal Society*, 151 (1861), pp. 431–577.
45. Russell, *History of Agricultural Science*; Burns and Hardy, *Nitrogen Fixation in Bacteria*; Hall, *Book of Rothamsted Experiments*.

5 Charles Darwin's Evolutionary Period

1. F. Darwin, 'The Botanical Work of Darwin', *Annals of Botany*, 13 (1899), pp. ix–xix; p. xi.
2. A. Sachs, *The Humbolt Current* (Oxford: Oxford University Press, 2007).
3. F. Darwin, 'Botanical Work of Darwin', p. xi.
4. Browne, *Charles Darwin: Voyaging*.
5. *Life and Letters of Charles Darwin*, vol. 3, p. 117.
6. King-Hele, *Life of Unequalled Achievement*, p. 218.
7. C. R. Darwin, *Autobiography of Charles Darwin 1809–1882* (1958), excerpted in *Works of Charles Darwin*, vol. 29, p. 76.
8. Ibid., vol. 29, p. 102; S. M. Walters and E. A. Stowe, *Darwin's Mentor: John Stevens Henslow, 1796–1861* (Cambridge: Cambridge University Press, 2001).
9. Walters and Stowe, *Darwin's Mentor*.
10. D. Kohn, G. Murrell, J. Parker and M. Whitehorn, 'What Henslow Taught Darwin', *Nature*, 436 (2005), pp. 643–5.

11. Shteir, *Cultivating Women*, p. 155.
12. W. T. Stearn, *John Lindley, 1799–1865: Gardener-botanist and Pioneer Orchidologist* (Woodbridge: Antique Collector's Club and Royal Horticultural Society; 1999).
13. C. R. Darwin, *Charles Darwin's Marginalia, Vol I*, ed. M. A. Di Gregorio and N. W. Gill (New York: Garland Press, 1990).
14. J. Schiller and T. Schiller, *Henri Dutrochet (1776–1847): le Materialisme Mécaniste at la Physiologie Générale* (Paris: Albert Blanchard, 1975).
15. Morton, *History of Botanical Science*; J. V. Pickstone, 'Locating Dutrochet', *British Journal of the History of Science*, 11 (1978), pp. 49–64.
16. J. S. Henslow, *The Principles of Descriptive and Physiological Botany* (London: Longmans, 1835), p. 157.
17. C. R. Darwin, *Autobiography*, excerpted in *Works of Charles Darwin*, vol. 29, p. 102.
18. Ibid., vol. 29, p. 105.
19. Walters and Stowe, *Darwin's Mentor*.
20. Ibid., p. 34.
21. Kohn, Murrell, Parker and Whitehorn, 'What Henslow Taught Darwin'.
22. D. M. Porter, 'Charles Darwin's Notes on Plants of the *Beagle* Voyage', *Taxon*, 31 (1982), pp. 503–6.
23. Browne, *Charles Darwin: Voyaging*; Walters and Stowe, *Darwin's Mentor*.
24. Browne, *Charles Darwin: Voyaging*.
25. C. R. Darwin, *On the Origin of Species*, in *Works of Charles Darwin*, vol. 15, p. 2.
26. Dupree, *Asa Gray*.
27. Cited by T. G. Hill, 'Presidential Address to Section K', *British Association for the Advancement of Science*, 20 (1931), pp. 2–20; p. 9.
28. F. Darwin, 'Botanical Work of Darwin', p. xii.
29. Charles Darwin to Asa Gray, 26 June 1863, letter 4222, *Darwin Correspondence Project*, www.darwinproject.ac.uk.
30. Charles Darwin to J. D. Hooker, 25 August 1863, letter 4274, *Darwin Correspondence Project*, www.darwinproject.ac.uk.
31. F. Darwin, 'Botanical Work of Darwin'.
32. J. Gilmour, *British Botanists* (London: William Collins, 1944).
33. C. R. Darwin, *Autobiography*, excerpted in *Works of Charles Darwin*, vol. 29, p. 130.
34. Ibid.
35. Veak, 'Exploring Darwin's Correspondence'.
36. R. Ornduff, 'Darwin's Botany', *Taxon*, 33 (1984), pp. 39–47.
37. F. Darwin, 'Botanical Work of Darwin'.
38. Cited in D. M. Simpkins, 'Knight, Thomas Andrew', in *Dictionary of Scientific Biography* (New York: Scribners, 1971), pp. 408–10.
39. Walters and Stowe, *Darwin's Mentor*; Kohn, Murrell, Parker and Whitehorn, 'What Henslow Taught Darwin'.
40. Browne, *Charles Darwin: The Power of Place*, p. 210.
41. C. R. Darwin, 'On the Existence of Two Forms, and on the Reciprocal Sexual Relation, in Several Species of the Genus Linum', *Journal of the Proceedings of the Linnean Society (Botany)*, 7 (1863), pp. 69–83; p. 77.
42. C. R. Darwin, *On the Various Contrivances by which British and Foreign Orchids are Fertilized by Insects [and the Good Effects of Intercrossing]* (London: John Murray, 1862), p. xiv.
43. Ibid., p. 257.

44. Ibid., p. 1.
45. C. R. Darwin, *The Origin of Species*, 6th edn, in *Works of Charles Darwin*, vol. 16, p. 162.

6 Charles Darwin's Physiological Period

1. F. Darwin, 'The Botanical Work of Darwin'.
2. J. H. F. Bothwell, 'Letters. The Long Past of Systems Biology', *New Phytologist*, 170 (2006), pp. 6–10.
3. D. C. Dennett, *Darwin's Dangerous Idea* (London: Allen Lane, 1955).
4. Bothwell, 'Letters'.
5. C. R. Darwin, *The Movements and Habits of Climbing Plants* (London: John Murray, 1865), p. 9.
6. J. G. Lennox, 'Darwin *Was* a Teleologist', *Biology & Philosophy*, 8 (1993), pp. 408–21.
7. C. R. Darwin, *Autobiography*, in *Works of Charles Darwin*, vol. 29, p. 152.
8. Browne, *Charles Darwin: The Power of Place*.
9. Von Sachs, *History of Botany*.
10. C. R. Darwin, *Climbing Plants*; Charles Darwin to J. D. Hooker, 13 August 1863, letter 4266, *Darwin Correspondence Project*, www.darwinproject.ac.uk.
11. Tendrils climbers may use modified shoots (as in the case of the grapevine, *Vitis*), or specialized leaves (pea, *Pisum)*, or leaflets (vetch, *Vicia*, or sweet pea, *Lathyrus)* or, in Passion Flower, *Passiflora gracilis,* tendrils derived from a flower stalk. Charles remarked, the tendril of Passiflora 'exceeds all other climbing plants ... in the rapidity of its movements'.
12. A fine glass filament, 5–20 mm long, and carrying a minute bead of black sealing wax, was attached to the organ tip. Below or behind the organ was fixed a white card with a black dot at its centre. Viewing the bead and the dot through a horizontal glass plate, a mark was made on the plate where they coincided. After regular, repeated observations the dots could be joined, to show the ellipses, and the pattern was copied onto tracing paper.
13. H. Opik and S. Rolfe, *The Physiology of Flowering Plants*, 4th edn (Cambridge: Cambridge University Press, 2005).
14. C. R. Darwin, *Climbing Plants*, p. 120.
15. Ibid.
16. Ibid., p. 122.
17. Ibid..
18. Ibid., p. 123.
19. Krikorian, 'Excerpts from the History of Plant Physiology and Development'.
20. Schiller and Schiller, *Henri Dutrochet*.
21. J. V. Pickstone, 'Discovering the Movement of Life: Osmosis and Microstructure in 1826', *Aspects of the History of Microcirculation*, 14 (1994), pp. 77–82; p. 81.
22 If short lengths of the flowering stem of a dandelion are split longitudinally, opposite halves of those placed in water will curl outwards as the inner cells take up water and increase in volume, while those placed a solution of salt will remain straight or even curl inwards as the inner cells of the stem lose water and shrink.
23. Wilhelm Pfeffer (1845–1920) was a student of Sachs; his own 256 students, 'Pepper Plants' [*Pfeffergewachse*], spread throughout the world. A lover of mountains, he was only the fifth man to climb the Matterhorn. Like Sachs, Pfeffer was prone to overwork

and depression in later life, in his case made infinitely worse by the loss of his son and only child in World War I. See Bünning, *Ahead of his Time*.

24. Ibid.; J. B. Sanderson, 'The Excitability of Plants I', *Nature*, 26 (1882), pp. 353–6.
25. Bunning, *Ahead of his Time*, p. 29.
26. J. Millett, R. I. Jones and S. Waldron, 'The Contribution of Insect Prey to the Total Nitrogen Content of Sundews (*Drosera* spp.) Determined in situ by Stable Isotope Analysis', *New Phytologist*, 158 (2003), pp. 527–34.
27. Browne, *Charles Darwin: Power of Place*.
28. M. Allan, *Darwin and his Flowers: The Key to Natural Selection* (London: Faber & Faber, 1977), p. 194.
29. Browne, *Charles Darwin: The Power of Place*, p. 149.
30. C. R. Darwin, *Insectivorous Plants* (London: John Murray, 1875), p. 1.
31. E. Darwin, *Loves of Plants*, canto I, ll. 229–40.
32. Ibid., p. 143.
33. Smith, *An Introduction to Physiological and Systematical Botany*, pp. 149, 150.
34. C. R. Darwin, *Insectivorous Plants*, 2nd edn, rev. by F. Darwin (1888), reprinted in *Works of Charles Darwin*, vol. 24, p. 1.
35. A. M. Ellison, N. J. Gotelli, J. S. Brewer, D. L. Cochran-Stafira, J. M. Kneitel, T. E. Miller, A. C. Worley and R. Zamora, 'The Evolutionary Ecology of Carnivorous Plants', *Advances in Ecological Research*, 33 (2003), pp. 1–74.
36. *Utricularia* spp. use suction traps to catch invertebrates. These aquatic plants have bladders, each of which is closed by a valve. Charles observed,

    > Animals enter the bladders by bending inwards the posterior free edge of the valve, which from being highly elastic shuts again instantly. As the edge is extremely thin, and fits closely against the edge of the collar, both projecting into the bladder, it would evidently be very difficult for any animal to get out when once imprisoned, and apparently they never do escape. To show how closely the edge fits, I may mention that my son found a Daphnia which had inserted one of its antennae into the slit, and it was thus held fast during a whole day.

    Charles believed that the bristles served to guide animals into the traps (C. R. Darwin, *Insectivorous Plants*, p. 405).
37. C. R. Darwin, *Insectivorous Plants*, p. 12.
38. Ibid., p. 11.
39. Ibid.
40. Ibid., p. 129.
41. Ibid., p. 10.
42. Ibid., p. 13.
43. N. Morgan, 'The Development of Biochemistry in England through Botany and the Brewing Industry (1870–1890)', *History and Philosophy of Sciences*, 2 (1980), pp. 141–66.
44. S. H. Vines, 'On the Digestive Ferment of *Nepenthes*', *Journal of the Linnean Society*, 15 (1877), pp. 427–31.
45. *Life and Letters of Charles Darwin*, vol. 3, p. 321.
46. C. R. Darwin, *Insectivorous Plants*, p. 42.
47. T. Romano, *Making Medicine Scientific: John Burdon Sanderson and the Culture of Victorian Science* (Baltimore, MD: Johns Hopkins University Press, 2002).

48. Ibid.
49. J. B. Sanderson and F. J. M. Page, 'On the Mechanical Effects and on the Electrical Disturbance Consequent on Excitation of the Leaf of *Dionaea muscipula*', *Proceedings of the Royal Society*, 25 (1876), pp. 411–34.
50. C. R. Darwin, *Insectivorous Plants*, 2nd edn, in *Works of Charles Darwin*, vol. 24, p. 237.
51. Sanderson, 'Excitability of Plants I'; J. B. Sanderson, 'The Excitability of Plants II', *Nature*, 26 (1882), pp. 482–6.
52. Sanderson, 'Excitability of Plants I', p. 356.
53. J. B. Sanderson, 'On the Electromotive Properties of the Leaf of Dioneae in the Excited and Unexcited States', *Philosophical Transactions of the Royal Society*, 173 (1888), pp. 1–55; p. 51.
54. Browne, *Charles Darwin: The Power of Place.*

7 Charles Darwin, Francis Darwin and Differences with von Sachs

1. Romano, *Making Medicine Scientific.*
2. See the website of the Brown Institute for Animals, http://www.vauxhallsociety.org.uk/Brown.html; F. Darwin, *Rustic Sounds and Other Studies in Literature and Natural History* (London: John Murray, 1917).
3. F. Darwin, 'Contribution to the Anatomy of the Sympathetic Ganglia of the Bladder in their Relation to the Vascular System', *Quarterly Journal of the Microscopical Society*, 14 (1874), pp. 109–14.
4. F. Darwin, 'On the Primary Vascular Dilatation in Acute Inflammation', *Journal of Anatomy and Physiology*, 10 (1876), pp. 1–16.
5. F. Darwin, 'On the Structure of the Snail's Heart', *Journal of Anatomy and Physiology*, 10 (1876), pp. 506–10.
6. G. L. Geison, *Michael Foster and the Cambridge School of Physiology* (Princeton, NJ: Princeton University Press, 1978).
7. Charles Darwin to Erasmus Darwin, 20 September 1873, letter 9060, *Darwin Correspondence Project*, www.darwinproject.ac.uk.
8. Browne, *Charles Darwin: The Power of Place.*
9. Sanderson, 'On the Electromotive Properties of the Leaf of Dioneae'.
10. F. Darwin 'The Process of Aggregation in the Tentacles of *Drosera rotundifolia*.' *Quarterly Journal of the Microscopical Society*, 16 (1876), pp. 309–19.
11. Browne, *Charles Darwin: The Power of Place.*
12. C. R. Darwin 'The Contractile Filaments of the Teasel', *Nature*, 16 (1877), p. 339.
13. Ibid.
14. Julius von Sachs (1832–97) was 'a sturdy figure of middle height, looking more like an artist than a professor. His rubicund face, with its reddish moustache and closely clipped pointed beard, was surmounted by a shock of darker hair brushed back from his rather prominent forehead, and his eyes sparkled with humour' (S. H. Vines, 'Reminiscences of German Botanical Laboratories in the "Seventies" and "Eighties" of the Last Century', *New Phytologist*, 24 (1925), pp. 1–8; p. 4). He regarded himself as old at forty-one and used stimulants, ranging from champagne to a mixture of camphor and chloralhydrate, before lecturing (W. O. James, 'Julius Sachs and the Nineteenth-Century Renaissance of Botany', *Endeavour*, 28 (1969), pp. 60–4).
15. Opik and Rolfe, *Physiology of Flowering Plants.*

16. K. Goebel. 'Julius Sachs', *Science Progress*, 7 (1898), pp. 150–73.
17. C. L. Dodgson, *The Vision of the Three Ts* (1873), reprinted in *The Complete Works of Lewis Carroll* (London: Nonesuch Library, 1988), p. 1040.
18. C. R. Darwin, *The Power of Movement in Plants*, assisted by F. Darwin (1880), in *Works of Charles Darwin*, vol. 27, p. 1.
19. Ibid.
20. T. A. Knight, 'On the Direction of the Germen and Radicle during the Vegetation of Seeds', *Philosophical Transactions of the Royal Society*, 96 (1806), pp. 99–108; pp. 100, 101.
21. C. R. Darwin, *Power of Movement*, in *Works of Charles Darwin*, vol. 27, pp. 418–19.
22. J. Heslop-Harrison, 'Darwin and the Movement of Plants: A Retrospect' in F. Skoog (ed.), *Plant Growth Substances 1979* (New York: Springer-Verlag, 1980), pp. 3–14.
23. J. von Sachs, *Vorlesungen uber Pflanzenphysiologie*, trans. H. M. Ward (Oxford: Clarendon Press, [1882] 1887), p. 689.
24. Cited in S. de Chadarevian, 'Laboratory Science versus Country-House Experiments. The Controversy between Julius Sachs and Charles Darwin' in *British Journal of the History of Science*, 29 (1996), pp. 17–41. There is a small possibility that Sachs was referring to Charles's earlier book, *On the Movements and Habits of Climbing Plants* (1875).
25. F. Darwin, 'Recollections', in *Springtime and Other Essays*, pp. 68, 69.
26. Opik and Rolfe, *Physiology of Flowering Plants*; J. Braam, 'In Touch: Plant Responses to Mechanical Stimuli' in *New Phytologist*, 165 (2005), pp. 373–89.
27. C. R. Darwin, *Power of Movement*, in *Works of Charles Darwin*, vol. 27, p. 3.
28. The rapid touch-induced closure of leaflets may protect them from herbivores by presenting a smaller surface area and exposing spines. In less than one second after the leaf is touched, motor cells can lose 25 per cent of their cell volume as water rushes out into the cell walls and spaces between cells. See Braam, 'In Touch'.
29. Heslop-Harrison, 'Darwin and the Movement of Plants', p. 9.
30. Heslop-Harrison, 'Darwin and the Movement of Plants'; de Chadarevian, 'Laboratory Science versus Country-House Experiments'.
31. Heslop-Harrison, 'Darwin and the Movement of Plants'.
32. De Chadarevian, 'Laboratory Science versus Country-House Experiments'.
33. F. O. Bower, 'English and German Botany in the Middle and Towards the End of Last Century', *New Phytologist*, 24 (1925), pp. 129–37; D. H. Scott 'German Reminiscences of the Early "Eighties"', *New Phytologist*, 24 (1925), pp. 9–16; Vines, 'Reminiscences of German Botanical Laboratories'.
34. Vines, 'Reminiscences of German Botanical Laboratories', pp. 2, 4.
35. Bower, 'English and German Botany', p. 132.
36. J. R. Moore, 'Charles Darwin Lies in Westminster Abbey', *Biological Journal of the Linnean Society*, 17 (1982), pp. 97–113.
37. F. Darwin, 'The Analogies of Plant and Animal Life', *Nature*, 17 (1878), pp. 411–14; p. 414.

8 Francis Darwin, Cambridge and Plant Physiology

1. Francis Darwin, letter to Nain Ruck, 20 December 1882. Darwin manuscript collection. Cambridge University Library, DAR 199.4 114.
2 Francis Darwin, letter to Nain Ruck, 11 March 1883. Darwin manuscript collection. Cambridge University Library, DAR 199.4 154.

3. W. Thiselton-Dyer, 'Plant Biology in the "Seventies"', *Nature*, 115 (1925), pp. 709–12.
4. C. Bibby, *T. H. Huxley: Scientist, Humanist and Educator* (London: Watts, 1959).
5. M. Argles, *South Kensington to Robbins: An Account of Technical and Scientific Education since 1851* (London: Longmans, 1964).
6. P. G. Ayres, *Harry Marshall Ward and the Fungal Thread of Death* (St. Paul, MN: American Phytopathological Society, 2005).
7. Bibby, *T. H. Huxley*, p. 111.
8. Ibid.
9. Geison, *Michael Foster*.
10. T. H. Huxley and H. N. Martin, *A Course of Practical Instruction in Elementary Biology* (London: Macmillan, 1875), p. ix.
11. J. R. Green, *A History of Botany in the United Kingdom from the Earliest Times to the End of the 19th Century* (London: J. M. Dent, 1914).
12. Ayres, *Harry Marshall Ward*.
13. F. O. Bower, *Sixty Years of Botany in Britain (1875–1935): Impressions of an Eyewitness* (London: Macmillan, 1938).
14. Vines, 'Reminiscences of German Botanical Laboratories', p. 3
15. Scott, 'German Reminiscences of the Early "Eighties"', p. 16
16. Green, *History of Botany*.
17. Ibid.; S. M. Walters, *The Shaping of Cambridge Botany* (Cambridge: Cambridge University Press, 1981).
18. Bower, *Sixty Years of Botany in Britain*.
19. Browne, *Charles Darwin: The Power of Place*.
20. S. Herbert and L. McKernan, *Who's Who in Victorian Cinema* (London: British Film Industry Publications, 1996).
21. Stephanie Jenkins, *pers. comm.*, and records of St Andrews church, Headington, Oxon.
22. Raverat, *Period Piece*; Spalding, *Gwen Raverat*.
23. A. C. Seward and F. F. Blackman, 'Obituary Notice. Francis Darwin (1848–1925)', *Proceedings of the Royal Society*, B, 110 (1932), pp. i–xxi.
24. Ibid.; F. Darwin, 'The Statolith-Theory of Geotropism', *Proceedings of the Royal Society*, 71 (1903), pp. 362–73; F. Darwin, 'Opening Address', *Nature*, 70 (1904), pp. 466–73.
25. K. Linsbauer, L. Linsbauer and L. R. Portheim (eds), *Wiesner und Seine Schule. Ein Beitrag zur Geschichte der Botanick. Festschrift* (Wien: n.p. 1903).
26. F. Darwin and R. W. Phillips, 'On the Transpiration Stream in Cut Branches', *Proceedings of the Cambridge Philosophical Society*, 5 (1884), p. 364.
27. Seward and Blackman, 'Obituary Notice'.
28. F. Darwin, 'Picturesque Experiments', in *Rustic Sounds*, pp. 200–18; p. 214.
29. F. Darwin, 'IX Observations on Stomata', *Philosophical Transactions of the Royal Society*, B, 190 (1898), pp. 561–621.
30. Frederick Frost Blackman FRS (1866–1947) was a knowledgeable member of the Fitzwilliam Museum Syndicate, about whom the Master of his college, St John's, said 'It was a great experience for a young Arts student to sit down at High Table with Bateson, Rivers [WHRR, the psychiatrist who pioneered treatment of shell-shock] and Blackman, and none of them was more friendly and impressive than Blackman' (G. E. Briggs, 'Frederick Frost Blackman, 1866–1947', *Obituary Notices of Fellows of the Royal Society*, 5 (1948), pp. 651–8; p. 625). A memorial biography written for the American Society of Plant Physiologists, of which Blackman was the first foreigner to be honoured as a Corresponding Member, records, 'F. F. Blackman, as no other man, influenced the trend of

the subject in the English speaking world' (F. C. Steward, 'In Memoriam. Frederick Frost Blackman', *Plant Physiology*, 22 (1947), pp. ii–viii; p. iv).

31. Briggs, 'Frederick Frost Blackman, 1866–1947'; [Anon.], 'Obituary. Sir Francis Darwin', *The Times*, Monday 21 September.
32. H. T. Brown and F. Escombe, 'Static Diffusion of Gases and Liquids in Relation to the Assimilation of Carbon and Translocation in Plants', *Philosophical Transactions of the Royal Society*, 193 (1900), pp. 223–91.
33. Ibid., p. 224.
34. Ibid., p. 278.
35. G. L. C. Matthaei, 'IV. Experimental Researches on Vegetable Assimilation and Respiration. 3. On the Effects of Temperature on Carbon-Dioxide Assimilation', *Philosophical Transactions of the Royal Society*, B, 197 (1904), pp. 47–105; F. F. Blackman and G. L. C. Matthaei, 'Experimental Researches on Vegetable Assimilation and Respiration. IV. A Quantitative Study of Carbon-Dioxide Assimilation and Leaf Temperature in Natural Illumination', *Proceedings of the Royal Society*, B, 76 (1905), pp. 402–60.
36. D. A. Walker, '"And Whose Bright Presence" – an Appreciation of Robert Hill and his Reaction', *Photosynthesis Research*, 73 (2002), pp. 51–4.
37. F. F. Blackman, 'Optima and Limiting Factors' in *Annals of Botany*, 19 (1905), pp. 281–95; p. 289.
38. Steward, 'In Memoriam. Frederick Frost Blackman', p. vi.
39. F. Darwin and D. F. M. Pertz, 'On a New Method of Estimating the Aperture of Stomata', *Proceedings of the Royal Society*, B, 84 (1911), pp. 136–54.
40. W. Detmer, *Das kleine Pflanzenphysiologische Praktikum*, 2nd edn, trans. S. Moor (London: Swan Sonnenschein & Co., 1898), p. vii; F. F. Blackman, 'X. Experimental Researches on Vegetable Assimilation and Respiration. No. 1. On a New Method for Investigating the Carbonic Exchange of Plants', *Philosophical Transactions of the Royal Society*, B, 186 (1895), pp. 485–502, p. vii.
41. Ibid., p. v.
42. F. Darwin and E. H. Acton, *Practical Physiology of Plants* (Cambridge: Cambridge University Press, 1894), p. v.
43. [Anon.], Obituary. Sir Francis Darwin', *The Times*, 21 September 1925, p. 14.
44. Walters, *Shaping of Cambridge Botany*.
45. W. Thiselton-Dyer, 'Harry Marshall Ward', *New Phytologist*, 6 (1907), pp. 1–9, p. 6.
46. Ayres, *Harry Marshall Ward*.
47. B. Darwin, *The World that Fred Made*.
48. Ibid., p. 38.
49. Ayres, *Harry Marshall Ward*.
50. A. D. Boney, 'The "Tansley Manifesto" affair', *New Phytologist*, 118 (1991), pp. 3–21.
51. F. F. Blackman, 'X. Experimental Researches on Vegetable Assimilation and Respiration. No. 1. On a New Method for Investigating the Carbonic Exchange of Plants', *Philosophical Transactions of the Royal Society*, B, 186 (1895), pp. 485–502.
52. O. V. S. Heath, *The Physiological Aspects of Photosynthesis* (London: Heinemann, 1969).

9 Francis Darwin, Family and his Father's Memory

1. Darwin, *Rustic Sounds*; Healey, *Emma Darwin*.
2. Browne, *Charles Darwin: The Power of Place*.
3 Death certificate, General Register Office.

4. Browne, *Charles Darwin: The Power of Place.*
5. Spalding, *Gwen Raverat.*
6. C. R. Darwin, *Autobiography*, in *Works of Charles Darwin*, vol. 29, pp. 118–19.
7. H. Litchfield (ed.), *Emma Darwin: A Century of Family Letters 1792–1896* (London: John Murray, 1915), p. 280.
8. [Anon.], 'Francis Darwin 1848–1925 Obituary', *Proceedings of the Royal Society*, 110 (1932), pp. i–xxi; p. iii.
9. Ibid.
10. Ayres, *Harry Marshall Ward.*
11. Spalding, *Gwen Raverat.*
12. Raverat, *Period Piece.*
13. F. Darwin, *The Story of a Childhood* (London: Oliver & Boyd, 1920).
14. B. Darwin, *The World that Fred Made*, p. 28.
15. N. Annan, *The Dons: Mentors, Eccentrics and Geniuses* (London: Harper Collins, 1999).
16. J. G. Stewart, *Jane Ellen Harrison* (London: Merlin Press, 1959), p. 21.
17. Ibid., pp. 102, 105.
18. Spalding, *Gwen Raverat.*
19. G. Johnson, *University Politics: F. M. Cornford's Cambridge and his Advice to the Young Academic Politician* (Cambridge: Cambridge University Press, 1994).
20. A. Bateson and F. Darwin, 'On a Method of Studying Geotropism', *Annals of Botany*, 2 (1888), pp. 65–8.
21. A. Bateson and F. Darwin, 'The Effect of Stimulation on Turgescent Vegetable Tissues', *Journal of the Linnean Society*, 24 (1888), pp. 1–27.
22. M. R. S. Creese, *Ladies in the Laboratory? American and British Women in Science, 1800–1900* (London: Scarecrow Press, 1998).
23. Ibid.; A. Arber, 'Miss Dorothea F. M. Pertz', *Nature*, 143 (1939), pp. 590–1.
24. M. L. Richmond, 'Women in the Early History of Genetics. William Bateson and the Newnham College Mendelians, 1900–1910', *Isis*, 92 (2001), pp. 55–90.
25. Ibid.
26. M. L. Richmond 'The "Domestication" of Heredity: the Familial Organisation of Geneticists at Cambridge University, 1895–1910', *Journal of the History of Biology*, 39 (2006), pp. 565–605.
27. M. L. Richmond, 'Muriel Wheldale Onslow and Early Biochemical Genetics', *Journal of the History of Biology*, 40 (2007), pp. 399–426.
28. Stewart, *Jane Ellen Harrison*, p. 105.
29. [Anon.], *Minutes of the General Board, Cambridge University*, 19 October 1904, Min. II.2.
30. Spalding, *Gwen Raverat.*
31. Stewart, *Jane Ellen Harrison*, p. 105.
32. H. Fowler, 'Frances Cornford, 1886–1960', in E. Shils and C. Blacker (eds), *Cambridge Women: Twelve Portraits* (Cambridge: Cambridge University Press, 1996), pp. 137–58.
33. F. MacCarthy, *Eric Gill* (London: Faber and Faber, 1989).
34. Spalding, *Gwen Raverat.*
35. W. Rothenstein, *Men and Memories: Recollections of William Rothenstein 1900–1922* (London: Faber and Faber, 1932), p. 32.
36. Ibid., p. 187.

37. Cited in E. Mayr, *The Growth of Biological Thought* (Harvard, MA: Belknap Press, 1982).
38. B. Bateson, *William Bateson, F.R.S. Naturalist: His Essays and Addresses* (Cambridge: Cambridge University Press, 1928).
39. Ibid.
40. Ibid.; Richmond, '"Domestication" of Heredity'.
41. P. Bateson, 'William Bateson: A Biologist Ahead of his Time', *Journal of Genetics*, 81 (2002), pp. 49–58; R. C. Punnet, 'Early Days of Genetics', *Heredity*, 4 (1950), pp. 1–10; J. H. Bennett, *Natural Selection, Heredity and Eugenics* (Oxford: Oxford University Press, 1983).
42. Bennett, *Natural Selection*, p. 95.
43. M. J. Chrispeels and D. E. Sadava, *Plants, Genes, and Crop Biotechnology* (London: Jones & Bartlett, 2003).
44. F. Darwin, 'The Butler/Darwin Controversy', in *Works of Charles Darwin*, vol. 29.
45. Ibid., p. 182
46. Ibid.
47. Ibid., pp. 176–210; p. 184.
48. Ibid., p. 194.
49. Ibid., p. 196.
50. F. Darwin, 'The Butler/Darwin Controversy', in *Works of Charles Darwin*, vol. 29.
51. F. Darwin and D. F. M. Pertz, 'On the Artificial Production of Rhythm in Plants', *Annals of Botany*, 6 (1892), pp. 245–64.
52. Ibid., p. 263
53. F. Darwin, 'Presidential Address' in *Report of the Seventy Eighth Meeting of the British Association for the Advancement of Science. Dublin 1908* (London: John Murray, 1909), pp. 3–27.
54. Ibid., p. 4.
55. Ibid., p. 10.
56. Ibid., p. 13.
57. Ibid.
58. Ibid., p. 14.
59. C. R. Darwin, *The Origin of Species*, 6th edn, in *Works of Charles Darwin*, vol. 16, p. 129.
60. F. Darwin, 'Presidential Address', p. 16.
61. Ibid., p. 17.
62 C. R. Darwin, *The Variation of Animals and Plants under Domestication* (London: John Murray, 1868).
63. [Anon.], 'Francis Darwin 1848–1925 Obituary'; Rothenstein, *Men and Memories.*
64. S. Morris, L. Wilson and D. Kohn, *Charles Darwin at Down House* (London: English Heritage, 1998).
65. C. R. Darwin, *Autobiography*, in *Works of Charles Darwin*, vol. 29, p. 144.
66. Rothenstein, *Men and Memories*, p. 156.
67. Raverat, *Period Piece*.
68. Rothenstein, *Men and Memories*; R. Speight, *William Rothenstein: The Portrait of an Artist in His Time* (London: Eyre and Spottiswood, 1962).
69. H. Godwin, *Cambridge and Clare* (Cambridge: Cambridge University Press, 1985), p. 57.

10 Fortune's Favourites?

1. S. Butler, *The Way of All Flesh* (London: Penguin Classics, [1903] 1973), p. 49.
2. Spalding, *Gwen Raverat*, p. 14.
3. Ibid.
4. Annan, *The Dons.*
5. Darwin, *The World that Fred Made.*
6. F. Cornford, *Selected Poems*. ed. Jane Dowson (London: Enitharman Press, 1996), p. 27.
7. Johnson, *University Politics*, p. 18.
8. M. Paley-Marshall, *What I Remember* (Cambridge: Cambridge University Press, 1947).
9. Spalding, *Gwen Raverat.*
10. Seward and Blackman, 'Francis Darwin 1848–1925. Obituary Notice'.
11. E. Darwin, 'The Economy of Vegetation', footnote to canto I, in *The Botanic Garden.*
12. E. Darwin, *The Temple of Nature*; or *The Origin of Society* (London: J Johnson, 1806), canto I, ll. 295–302.
13. Ibid., canto IV, ll. 42–5.
14. C. R. Darwin, *The Origin of Species*, 6th edn, in *Works of Charles Darwin*, vol. 16, p. 000.
15. King-Hele, *Erasmus Darwin.*
16. C. R. Darwin, *Life of Erasmus Darwin*, in *Works of Charles Darwin*, vol. 29, p. xiv.
17. King-Hele, *Erasmus Darwin*, p. 84.
18. Ibid.
19. King-Hele, *Life of Unequalled Achievement*, p. 366.
20. Ibid.
21. Ibid.
22. E. Darwin, *Phytologia*, p. 26.
23. King-Hele, *Erasmus Darwin*, p. 88.
24. Charles Darwin to J. D. Hooker, 5 June 1855, letter 1693, *Darwin Correspondence Project*, www.darwinproject.ac.uk.
25. C. R. Darwin, *On the Origin of Species*, in *Works of Charles Darwin*, vol. 15, p. 257.
26. E. Darwin, *The Botanic Garden*, canto III.
27. D. Coffey 'Protecting the Botanic Garden: Seward, Darwin, and Coalbrookdale', *Women's Studies*, 31 (2002), pp. 141–65; Uglow, *Lunar Men.*
28. E. Darwin, *The Botanic Garden*, canto III.
29. Darwin, *Insectivorous Plants*, p. 22.
30. Ibid.
31. Browne, *Charles Darwin: The Power of Place.*
32. *Life and Letters of Charles Darwin*, vol. 1, p. 145.
33. Ibid., vol. 1, p. 155.
34. Ibid., vol. 1, p. 146.
35. Ibid., vol. 1, p. 148.
36. Ibid.
37. Ibid., vol. 1, p. 146.
38. Bateson, *William Bateson*; Richmond, 'The "Domestication" of Heredity'.
39. F. Darwin, 'The Nutrition of Drosera Rotundifolia', *Nature*, 18 (1878), pp. 153–4.
40. C. R. Darwin, *Autobiography*, p. 25.

41. F. Darwin, 'Darwin's Work on the Movement of Plants', in A. C. Seward (ed.), *Darwin and Modern Science: Essays in Commemoration of the Centenary of the Birth of Charles Darwin* (Cambridge: Cambridge University Press, 1909), pp. 385–400.
42. Spalding, *Gwen Raverat*, p. 14.

11 Where Did the Green Threads Lead? The Botanical Legacy

1. Allan, *Darwin and his Flowers.*
2. J. C. McKusick, 'Nature', in M. Ferber (ed.), *A Companion to European Romanticism* (Oxford: Blackwell, 2005), pp.413–32.
3. J. D Hooker to C. R. Darwin, 12 October 1862, letter 3757, *Darwin Correspondence Project*, www.darwinproject.ac.uk.
4. H. T. Brown and G. H. Morris, 'Researches on the Germination of Some of the Gramineae', *Journal of the Chemical Society*, 63 (1890), pp. 604–77.
5. H. T. Brown and G. H. Morris, 'A Contribution to the Chemistry and Physiology of Foliage Leaves', *Journal of the Chemical Society*, 57 (1893), pp. 458–528.
6. J. R. Green, *A History of Botany 1860–1900. Being a Continuation of Sachs' History of Botany, 1530–1860* (Oxford: Clarendon Press, 1909); Green, *A History of Botany in the United Kingdom.*
7. Green, *A History of Botany in the United Kingdom*; Morgan, 'The Development of Biochemistry'.
8. *Life and Letters of Charles Darwin.*
9. A. R. Mast and E. Conti, 'Commentary. The Primrose Path to Heterostyly', *New Phytologist*, 171 (2006), pp. 439–42.
10. Ibid.
11. F. W. Went, 'Reflections and Speculations', *Annual Review of Plant Physiology*, 25 (1974), pp. 1–26; p. 25.
12. Ibid.
13. Bunning, *Ahead of his Time.*
14. Went, 'Reflections and Speculations', p. 5.
15. K. V. Thimann, 'Plant Growth Substances, Past, Present and Future', *Annual Review of Plant Physiology*, 14 (1963), pp. 1–19.
16. Opik and Rolfe, *The Physiology of Flowering Plants.*
17. K. V. Thimann, 'The Auxins', in M. B. Wilkins (ed.), *The Physiology of Plant Growth and Development* (London: McGraw-Hill, 1969), pp. 2–45.
18. Thimann, 'Plant Growth Substances'.
19. D. Ingram, D. Vince-Prue and P. Gregory, *Science and the Garden: The Scientific Basis of Horticultural Practice* (Oxford: Blackwell, 2002); Taiz and Zeiger, *Plant Physiology.*
20. F. B. Salisbury and C. W. Ross, *Plant Physiology*, 4th edn (Belmont CA: Wadsworth, 1992).
21. H. G. Jones, 'Stomatal Control of Photosynthesis and Transpiration', *Journal of Experimental Botany*, 49 (1998), pp. 387–98.
22. B. R. Loveys, M. Stoll and W. J. Davies, 'Physiologic Approaches to Enhance Water use Efficiency in Agriculture: Exploiting Plant Signalling in vovel Irrigation Practice', in M. A. Bacon (ed.) *Water Use Efficiency in Plant Biology* (Cambridge: Cambridge University Press, 2004), pp. 113–41.
23. E. J. Maskell, 'Experimental Researches on Vegetable Assimilation and Respiration. XVIII. The Relation between Stomatal Opening and Assimilation - a Critical Study of

Assimilation Rates and Porometer Rates of Cherry Laurel', *Proceedings of the Royal Society*, B, 102 (1928), pp. 488–533.

24. A. S. D. Eller and J. P. Sparks, 'Predicting Leaf-Level Fluxes of O_3 and NO_2: The Relative Roles of Diffusion and Biochemical Processes', *Plant, Cell & Environment*, 29 (2006), pp. 1742–50.
25. A. M. Hetherington and F. I. Woodward, 'The Role of Stomata in Sensing and Driving Environmental Change', *Nature*, 424 (2003), pp. 901–8.
26. M. A. Bacon, 'Water use Efficiency in Plant Biology' in Bacon (ed.), *Water Use Efficiency in Plant Biology*, pp. 10–16.
27. F. I. Woodward, 'Stomatal Numbers are Sensitive to CO_2 Increases from Pre-Industrial Levels', *Nature*, 327 (1987), pp. 617–18.
28. Beerling, *Emerald Planet.*
29. [Anon.], 'Rowland Harry Biffen (1874–1949)', *Obituary Notices of Fellows of the Royal Society of London*, 7 (1950), pp. 9–25.
30. F. Kidd, C. West and G. E. Briggs, 'What is the Significance of the Efficiency Index of Plant Growth?', *New Phytologist*, 19 (1920), pp. 97–100.
31. Green, *History of Botany 1860–1900*; Green, *History of Botany in the United Kingdom.*
32. Bunning, *Ahead of his Time.*
33. Boney, 'The "Tansley Manifesto" Affair'.
34. F. F. Blackman, F. W. Oliver, V.H. Blackman, and F. Keeble, 'The Reconstruction of Elementary Botanical Teaching', *New Phytologist*, 16 (1917), pp. 241–52; p. 242.
35. Ibid.
36. Boney, 'The "Tansley Manifesto" Affair', p. 13.
37. A 60ft long canvas painted by William Rothenstein in 1916 hangs in the Senate Chamber of the University of Southampton. One of the earliest tributes to the war dead, its figures represent Vice-Chancellors, scholars, and leading men of science (including Francis Darwin) who surround a Chancellor conferring a degree upon a young soldier. Walking forward hand-in-hand to receive symbolically what could never be given to them is a group of undergraduates - all dead - which includes the poet Rupert Brooke, and Raymond Asquith (the Prime Minister's son). Francis Darwin is flanked by Edward Poulton (Professor of Zoology, University of Oxford) and Oliver Lodge (physicist and Vice-Chancellor, University of Birmingham).
38. E. D. Rudolph, 'History of the Botanical Teaching Laboratory in the United States' in *American Journal of Botany*, 83 (1996), pp. 661–71.
39. A. M. Hetherington and H. Slater, Editorial, 'Physiology & Development' in *New Phytologist*, 173 (2007), pp. 1–2; p. 1.
40. J. Small, 'The Student as a Synthesizing Organism', *New Phytologist*, 17 (1918), pp. 189–93; p. 190.

WORKS CITED

[Anon.], *Minutes of the General Board, Cambridge University*, 19 October 1904, Min. II.2.

—, 'Obituary. Sir Francis Darwin', *The Times*, 21 September 1925.

—, 'Francis Darwin 1848–1925 Obituary', *Proceedings of the Royal Society*, 110 (1932), pp. i–xxi.

—, 'Rowland Harry Biffen (1874–1949)', *Obituary Notices of Fellows of the Royal Society of London*, 7 (1950), pp. 9–25.

Ainsworth, E. A., and S. P. Long, 'What Have We Learned from 15 years of Free-Air CO_2 Enrichment (FACE). A Meta-Analytic Review of the Responses of Photosynthesis, Canopy Properties and Plant Production to Rising CO_2', *New Phytologist*, 165 (2005), pp. 351–72.

Allan D. C. G., and R. E. Schofield, *Stephen Hales: Scientist and Philanthropist* (London: Scholar Press, 1980).

Allan, M., *Darwin and His Flowers: The Key to Natural Selection* (London: Faber & Faber, 1977).

Annan, N., *The Dons: Mentors, Eccentrics and Geniuses* (London: Harper Collins, 1999).

Arber, A., 'Miss Dorothea F. M. Pertz', *Nature*, 143 (1939), pp. 590–1.

Argles, M., *South Kensington to Robbins: An Account of Technical and Scientific Education since 1851* (London: Longmans, 1964).

Ayres, P. G., *Harry Marshall Ward and the Fungal Thread of Death* (St Paul, MN: American Phytopathological Society, 2005).

Bacon, M. A., 'Water Use Efficiency in Plant Biology', in *Water Use Efficiency in Plant Biology* (Cambridge: Cambridge University Press, 2004), pp. 10–16.

Bateson, A., and F. Darwin, 'The Effect of Stimulation on Turgescent Vegetable Tissues', *Journal of the Linnean Society*, 24 (1888), pp. 1–27.

—, 'On a Method of Studying Geotropism', *Annals of Botany*, 2 (1888), pp. 65–8.

Bateson, B., *William Bateson, F.R.S. Naturalist: His Essays and Addresses* (Cambridge: Cambridge University Press, 1928).

Bateson, P., 'William Bateson: A Biologist Ahead of his Time', *Journal of Genetics*, 81 (2002), pp. 49–58.

Bateson, W., *Mendel's Principles of Heredity: A Defence* (Cambridge: Cambridge University Press, 1902).

—, *Mendel's Principles of Heredity* (Cambridge: Cambridge University Press, 1909).

Beale, N., and E. Beale, *Who was Ingen Housz, Anyway?: A Lost Genius* (Calne: Calne Town Council, 1999).

Beerling, D., *The Emerald Planet: How Plants Changed Earth's History* (Oxford: Oxford University Press, 2007).

Bennett, J. H., *Natural Selection, Heredity and Eugenics* (Oxford: Oxford University Press, 1983).

Bibby, C., *T. H. Huxley: Scientist, Humanist and Educator* (London: Watts, 1959).

Blackman, F. F., 'X. Experimental Researches on Vegetable Assimilation and Respiration. No. 1. On a New Method for Investigating the Carbonic Exchange of Plants', *Philosophical Transactions of the Royal Society*, B, 186 (1895), pp. 485–502.

—, 'Optima and Limiting Factors', *Annals of Botany*, 19 (1905), pp. 281–95.

Blackman, F. F., and G. L. C. Matthaei, 'Experimental Researches on Vegetable Assimilation and Respiration. IV. A Quantitative Study of Carbon-Dioxide Assimilation and Leaf Temperature in Natural Illumination', *Proceedings of the Royal Society*, B, 76 (1905), pp. 402–60.

Blackman, F. F., F. W. Oliver, V.H. Blackman, and F. Keeble, 'The Reconstruction of Elementary Botanical Teaching', *New Phytologist*, 16 (1917), pp. 241–52.

Boney, A. D., 'The "Tansley Manifesto" Affair', *New Phytologist*, 118 (1991), pp. 3–21.

Bothwell, J. H. F., 'Letters. The Long Past of Systems Biology', *New Phytologist*, 170, (2006), pp. 6–10.

Bower, F. O., 'English and German Botany in the Middle and Towards the End of Last Century', *New Phytologist*, 24 (1925), pp. 129–37.

—, *Sixty Years of Botany in Britain (1875–1935): Impressions of an Eyewitness* (London: Macmillan, 1938).

Bower, F. O., and S. H. Vines, *A Course of Practical Instruction in Botany, Part I* (London: Macmillan, 1885).

Braam, J., 'In Touch: Plant Responses to Mechanical Stimuli', *New Phytologist*, 165 (2005), pp. 373–89.

Briggs, A., 'The Later Victorian Age', in B. Ford (ed.), *The Cambridge Cultural History of Britain* (Cambridge: Cambridge University Press, 1995), pp. 2–78.

Briggs, G. E., 'Frederick Frost Blackman, 1866–1947', *Obituary Notices of Fellows of the Royal Society*, 5 (1948), pp. 651–8.

Brown, H. T., and F. Escombe, 'Static Diffusion of Gases and Liquids in Relation to the Assimilation of Carbon and Translocation in Plants', *Philosophical Transactions of the Royal Society*, 193 (1900), pp. 223–91.

Brown, H. T., and G. H. Morris, 'Researches on the Germination of Some of the Gramineae', *Journal of the Chemical Society*, 63 (1890), pp. 604–77.

—, 'A Contribution to the Chemistry and Physiology of Foliage Leaves', *Journal of the Chemical Society*, 57 (1893), pp. 458–528.

Browne, J., *Charles Darwin: Voyaging. Volume I of a Biography* (London: Pimlico, 1995).

—, *Charles Darwin: The Power of Place. Volume II of a Biography*. (London: Jonathan Cape, 2002).

Bunning, E., *Ahead of his Time: Wilhelm Pfeffer: Early Advances in Plant Biology*, trans. H. W. Pfeffer (Ottawa, ON: Carleton University Press, 1989).

Burns R. C., and W. F. R. Hardy, *Nitrogen Fixation in Bacteria and Higher Plants* (Berlin: Springer-Verlag, 1975).

Butler, S., *Evolution Old and New, or the Theories of Buffon, Dr Erasmus Darwin and Lamark Compared with that of Mr C. Darwin* (London: Hardwicke & Bogue, 1879).

—, *The Way of All Flesh* (London: Penguin Classics, [1903] 1973).

Chrispeels, M. J., and D. E. Sadava, *Plants, Genes, and Crop Biotechnology* (London: Jones & Bartlett, 2003).

Clark-Kennedy, A. E., *Stephen Hales, D.D., F.R.S.* (Cambridge: Cambridge University Press, 1929).

Cohen, I. B., 'Stephen Hales', *Scientific American*, 234 (1976), pp. 98–107.

Cornford, F., *Selected Poems*. ed. Jane Dowson (London: Enitharman Press, 1996).

Creese, M. R. S., *Ladies in the Laboratory? American and British Women in Science, 1800–1900* (London: Scarecrow Press, 1998).

Darwin, B., *The World that Fred Made* (London: Chatto and Windus, 1955).

Darwin, C. R, *On the Origin of Species by Means of Natural Selection, or the Preservation of Favoured Races in the Struggle for Life* (1859), reprinted in *The Works of Charles Darwin*, ed. P. H. Barrett and R. B. Freeman, 29 vols (London: Pickering & Chatto, 1988–9), vol. 15.

—, *On the Various Contrivances by which British and Foreign Orchids are Fertilized by Insects [and the Good Effects of Intercrossing]* (London: John Murray, 1862).

—, 'On the Existence of Two Forms, and on the Reciprocal Sexual Relation, in Several Species of the Genus Linum', *Journal of the Proceedings of the Linnean Society (Botany)*, 7 (1863), pp. 69–83.

—, *The Movements and Habits of Climbing Plants* (London: John Murray, 1865).

—, *The Variation of Animals and Plants under Domestication* (London: John Murray, 1868).

—, *The Movements and Habits of Climbing Plants*, 2nd, rev. edn (London: John Murray, 1875), reprinted in *The Works of Charles Darwin*, ed. P. H. Barrett and R. B. Freeman, 29 vols (London: Pickering & Chatto, 1988–9), vol. 18.

—, *The Effects of Cross and Self-fertilisation in the Vegetable Kingdom* (London: John Murray, 1876).

—, *The Origin of Species by Means of Natural Selection, or the Preservation of Favoured Races in the Struggle for Life*, 6th edn (1876), reprinted in *The Works of Charles Darwin*, ed. P. H. Barrett and R. B. Freeman, 29 vols (London: Pickering & Chatto, 1988–9), vol. 16.

—, 'The Contractile Filaments of the Teasel', *Nature*, 16 (1877), p. 339.

—, *The Different Forms of Flowers on Plants of the Same Species* (London: John Murray, 1877).

—, *The Various Contrivances by which Orchids Are Fertilised by Insects*; 2nd edn (1877), reprinted in *The Works of Charles Darwin*, ed. P. H. Barrett and R. B. Freeman, 29 vols (London: Pickering & Chatto, 1988–9), vol. 17.

—, *The Power of Movements in Plants*, assisted by F. Darwin (London: John Murray, 1880), reprinted in *The Works of Charles Darwin*, ed. P. H. Barrett and R. B. Freeman, 29 vols (London: Pickering & Chatto, 1988–9), vol. 27.

—, *The Movements and Habits of Climbing Plants*, 2nd edn (1882), reprinted in *The Works of Charles Darwin*, ed. P. H. Barrett and R. B. Freeman, 29 vols (London: Pickering & Chatto, 1988–9), vol. 18.

—, *The Life and Letters of Charles Darwin*, ed. F. Darwin, 3 vols (London: John Murray, 1887).

—, *Insectivorous Plants*, 2nd edn, rev. by F. Darwin (1888), reprinted in *The Works of Charles Darwin*, ed. P. H. Barrett and R. B. Freeman, 29 vols (London: Pickering & Chatto, 1988–9), p. 1.

—, *More Letters of Charles Darwin*, ed. F. Darwin and A. C. Seward (London: John Murray, 1903).

—, *The Autobiography of Charles Darwin 1809–82* (New York: Barnes and Noble Books, 1958).

—, *Darwin Correspondence Project* (1974–), www.darwinproject.ac.uk.

—, *The Autobiography of Charles Darwin 1809–82* (1958), excerpted in *The Works of Charles Darwin*, ed. P. H. Barrett and R. B. Freeman, 29 vols (London: Pickering & Chatto, 1988–9), vol. 29.

—, *The Works of Charles Darwin*, ed. P. H. Barrett and R. B. Freeman, 29 vols (London: Pickering & Chatto, 1989–90).

—, *Charles Darwin's Marginalia, Vol I*, ed. M. A. Di Gregorio and N. W. Gill, (New York: Garland Press, 1990).

Darwin, E., *Phytologia or the Philosophy of Agriculture and Gardening with the Theory of Draining Morasses and with an Improved Construction of the Drillplough* (London: J. Johnson, 1800).

—, *The Temple of Nature*; or *The Origin of Society* (London: J. Johnson, 1806).

—, *The Botanic Garden. A Poem in Two Parts; Containing The Economy of Vegetation and the Loves of The Plants* (London: Jones & Co., 1825).

—, *The Loves of the Plants; A Poem with Philosophical Notes* (London: Jones & Co., 1825).

Darwin, F., 'Contribution to the Anatomy of the Sympathetic Ganglia of the Bladder in their Relation to the Vascular System', *Quarterly Journal of the Microscopical Society*, 14 (1874), pp. 109–14.

—, 'On the Primary Vascular Dilatation in Acute Inflammation', *Journal of Anatomy and Physiology*, 10 (1876), pp. 1–16.

—, 'On the Structure of the Snail's Heart', *Journal of Anatomy and Physiology*, 10 (1876), pp. 506–10

—, 'The Process of Aggregation in the Tentacles of *Drosera rotundifolia*', *Quarterly Journal of the Microscopical Society*, 16 (1876), pp. 309–19.

—, 'The Analogies of Plant and Animal Life', *Nature*, 17 (1878), pp. 411–14.

—, 'The Nutrition of Drosera Rotundifolia', *Nature*, 18 (1878), pp. 153–4

—, 'Reminiscences of my Father's Everyday Life' (1884), in C. R. Darwin, *The Autobiography of Charles Darwin 1809–82* (New York: Barnes and Noble Books, 1958)

—, 'IX Observations on Stomata', *Philosophical Transactions of the Royal Society*, B, 190 (1898), pp. 561–621.

—, 'The Botanical Work of Darwin', *Annals of Botany*, 13 (1899), pp. x–xix.

—, 'The Statolith-Theory of Geotropism', *Proceedings of the Royal Society*, 71 (1903), pp. 362–73.

—, 'Opening Address', *Nature*, 70 (1904), pp. 466–73.

—, 'Darwin's Work on the Movement of Plants', in A. C. Seward (ed.), *Darwin and Modern Science: Essays in Commemoration of the Centenary of the Birth of Charles Darwin* (Cambridge: Cambridge University Press, 1909), pp. 385–400.

—, 'Presidential Address', *Report of the Seventy Eighth Meeting of the British Association for the Advancement of Science. Dublin 1908* (London: John Murray, 1909), pp. 3–27.

—, *Rustic Sounds and Other Studies in Literature and Natural History* (London: John Murray, 1917).

—, *Springtime and Other Essays* (London: John Murray, 1920), pp. 51–71.

—, *The Story of a Childhood* (London: Oliver & Boyd, 1920).

Darwin, F., and E. H. Acton, *Practical Physiology of Plants* (Cambridge: Cambridge University Press, 1894).

Darwin, F., and D. F. M. Pertz, 'On the Artificial Production of Rhythm in Plants', *Annals of Botany*, 6 (1892), pp. 245–64.

—, 'On a New Method of Estimating the Aperture of Stomata', *Proceedings of the Royal Society*, B, 84 (1911), pp. 136–54.

Darwin, F., and R. W. Phillips, 'On the Transpiration Stream in Cut Branches', *Proceedings of the Cambridge Philosophical Society*, 5 (1884), p. 364.

Davy, Sir H., *Elements of Agricultural Chemistry* (London: W. Bulmer and Co., 1813).

de Candolle, A. P., *Physiologie Végétale* (Paris: Béchet, 1832).

de Chadarevian, S., 'Laboratory Science versus Country-House Experiments. The Controversy between Julius Sachs and Charles Darwin', *British Journal of the History of Science*, 29 (1996), pp. 17–41.

de Saussure, N. T., *Recherches Chimiques sur la Végétation* (Paris: Didot jeune, 1804).

Dennett, D. C., *Darwin's Dangerous Idea* (London: Allen Lane, 1955).

Detmer, W., *Das kleine Pflanzenphysiologische Praktikum*, 2nd edn, trans. S. Moore (London: Swan Sonnenschein & Co., 1898).

Dodgson, C. L., *The Complete Works of Lewis Carroll* (London: Nonesuch Library, 1988).

Dupree, A. H., *Asa Gray 1810–1888* (Cambridge, MA: Harvard University Press, 1959).

Eller, A. S. D., and J. P. Sparks, 'Predicting Leaf-Level Fluxes of O_3 and NO_2: The Relative Roles of Diffusion and Biochemical Processes', *Plant, Cell & Environment*, 29 (2006), pp. 1742–50.

Ellison, A. M., N. J. Gotelli, J. S. Brewer, D. L. Cochran-Stafira, J. M. Kneitel, T. E. Miller, A. C. Worley and R. Zamora, 'The Evolutionary Ecology of Carnivorous Plants', *Advances in Ecological Research*, 33 (2003), pp. 1–74.

Farley, J., *Gametes and Spores* (Baltimore, MD: Johns Hopkins University Press, 1982).

Fowler, H., 'Frances Cornford, 1886–1960', in E. Shils and C. Blacker (eds), *Cambridge Women: Twelve Portraits* (Cambridge: Cambridge University Press, 1996), pp. 137–58.

Freeman, R. B., 'The Darwin Family', *Biological Journal of the Linnean Society*, 17, (1982), pp. 9–21.

Geison, G. L., *Michael Foster and the Cambridge School of Physiology* (Princeton, NJ: Princeton University Press, 1978).

Gest, H., 'A "Misplaced Chapter" in the *History of Photosynthesis Research*; the Second Publication (1796) on Plant Processes by Dr Jan Ingen-Housz, MD, Discoverer of Photosynthesis', *Photosynthesis Research*, 53 (1997), pp. 65–72.

—, 'Bicentenary Homage to Dr Jan Ingen-Housz, MD (1730–1799), Pioneer of Photosynthesis Research', *Photosynthesis Research*, 63 (2000), pp. 183–90.

Gibbs, F. W., *Joseph Priestley: Adventurer in Science and Champion of Truth* (London: Nelson, 1965).

Gilmour, J., *British Botanists* (London: William Collins, 1944).

Godwin, H., *Cambridge and Clare* (Cambridge: Cambridge University Press, 1985).

Goebel. K., 'Julius Sachs', *Science Progress*, 7 (1898), pp. 150–73.

Green, J. R., *A History of Botany 1860–1900. Being a Continuation of Sachs' History of Botany, 1530–1860* (Oxford: Clarendon Press, 1909).

—, *A History of Botany in the United Kingdom from the Earliest Times to the End of the 19th Century* (London: J. M. Dent, 1914).

Hales, S., *Vegetable Staticks*, reprinted from the 1st edn with a foreword by M. A. Hoskin (London: The Scientific Book Guild, [1727] 1961).

Hall, A. D., *The Book of Rothamsted Experiments* (London: John Murray, 1905).

Hart, H., 'Nicolas Theodore de Saussure', *Plant Physiology*, 5 (1930), pp. 424–9.

Healey, E., *Emma Darwin: The Inspirational Wife of a Genius* (London: Hodder Headline, 2001).

Heath, O. V. S., *The Physiological Aspects of Photosynthesis* (London: Heinemann, 1969).

Henslow, J. S., *The Principles of Descriptive and Physiological Botany* (London: Longmans, 1835).

Herbert, S., and L. McKernan, *Who's Who in Victorian Cinema* (London: British Film Industry Publications, 1996).

Heslop-Harrison, J., 'Darwin and the Movement of Plants: A Retrospect' in F. Skoog (ed.), *Plant Growth Substances 1979* (New York, NY: Springer-Verlag, 1980), pp. 3–14.

Hetherington, A. M., and H. Slater, Editorial, 'Physiology & Development', *New Phytologist*, 173 (2007), pp. 1–2.

Hetherington, A. M., and F. I. Woodward, 'The Role of Stomata in Sensing and Driving Environmental Change', *Nature*, 424 (2003), pp. 901–8.

Hill, T. G., 'Presidential Address to Section K', *British Association for the Advancement of Science*, 20 (1931), pp. 2–20.

Hooker, W. J., *The British Flora: Vol. I. Comprising the Phænogamous or Flowering Plants, and the Ferns* (London: Longman, 1842).

Huxley, T. H., and H. N. Martin, *A Course of Practical Instruction in Elementary Biology* (London: MacMillan, 1875).

Ingen-Housz, J., *Experiments upon Vegetables.* (London: Elmsly & Payne, 1779).

Ingram, D., D. Vince-Prue and P. Gregory, *Science and the Garden: The Scientific Basis of Horticultural Practice* (Oxford: Blackwell, 2002).

James, W. O., 'Julius Sachs and the Nineteenth-Century Renaissance of Botany', *Endeavour*, 28 (1969), pp. 60–4

Johnson, G., *University Politics. F. M. Cornford's Cambridge and his Advice to the Young Academic Politician* (Cambridge: Cambridge University Press, 1994).

Jones, H. G., 'Stomatal Control of Photosynthesis and Transpiration', *Journal of Experimental Botany*, 49 (1998), pp. 387–98.

Kidd, F., C. West and G. E. Briggs, 'What is the Significance of the Efficiency Index of Plant Growth?', *New Phytologist*, 19 (1920), pp. 97–100.

King-Hele, D., *Erasmus Darwin* (London: Macmillan, 1963).

—, *Erasmus Darwin: A Life of Unequalled Achievement* (London: De La Mare, 1998).

—, 'The 1997 Wilkins Lecture: Erasmus Darwin, the Lunaticks and evolution', *Notes and Records of the Royal Society of London*, 52 (1998), pp. 153–80.

Knight, T. A., 'On the Direction of the Germen and Radicle during the Vegetation of Seeds', *Philosophical Transactions of the Royal Society*, 96 (1806), pp. 98–101.

Kohn, D., G. Murrell, J. Parker and M. Whitehorn, 'What Henslow Taught Darwin', *Nature*, 436 (2005), pp. 643–5.

Körner, C., 'Plant CO_2 Responses: An Issue of Definition, Time and Resource Supply', *New Phytologist*, 172 (2006), pp. 393–411

Krause, E., *Erasmus Darwin ... With a Preliminary Notice by Charles Darwin* (1879), excerpted in *The Works of Charles Darwin*, ed. P. H. Barrett and R. B. Freeman, 29 vols (London: Pickering & Chatto, 1989), vol. 29, pp.

Krikorian, A. D., 'Excerpts from the History of Plant Physiology and Development', in P. J. Davies (ed.), *Historical and Current Aspects of Plant Physiology: A Symposium Honoring F. C. Steward* (New York: New York State University, Cornell, 1975), pp. 9–97.

Lankester, E., Review of *An Elementary Course in Botany* by A. Henfrey, *Athenaeum* (3 October 1857), p. 1562.

Lawes, J. B., J. H. Gilbert and E. Pugh, 'On the Sources of the Nitrogen of Vegetation; with Special Reference to the Question Whether Plants Assimilate Free or Uncombined Nitrogen', *Philosophical Transactions of the Royal Society*, 151 (1861), pp. 431–577.

Lennox, J. G., 'Darwin *Was* a Teleologist', *Biology & Philosophy*, 8 (1993), pp. 408–21.

Linsbauer, K., L. Linsbauer and L. R. Portheim (eds), *Wiesner und Seine Schule. Ein Beitrag zur Geschichte der Botanick. Festschrift* (Wien: n.p. 1903).

Litchfield, H., *Emma Darwin: A Century of Family Letters 1792–1896* (London: John Murray, 1915).

Loveys, B. R., M. Stoll and W. J. Davies, 'Physiologic Approaches to Enhance Water Use Efficiency in Agriculture: Exploiting Plant Signalling in Vovel Irrigation Practice', in M. A. Bacon (ed.), *Water Use Efficiency in Plant Biology*, (Cambridge: Cambridge University Press, 2004), pp. 113–41.

Lyte, C., *The Plant Hunters* (London: Orbis, 1983).

MacCarthy, F., *Eric Gill* (London: Faber and Faber, 1989).

McCosh, F. W. J., *Boussingault: Chemist and Agriculturalist* (Dordrecht: Kluwer, 1984).

Mackenzie, J. M., 'Plant Collecting and Imperialism', in J. Illingworth and J. Routh (eds.), *Reginald Farrar: Dalesman, Planthunter, Gardener* (Lancaster: Centre for North-West Regional Studies, University of Lancaster, 1991), pp. 8–14.

McKusick, J. C., 'Nature', in M. Ferber (ed.), *A Companion to European Romanticism* (Oxford: Blackwell, 2005), pp. 413–32.

Maskell, E. J., 'Experimental Researches on Vegetable Assimilation and Respiration. XVIII. The Relation between Stomatal Opening and Assimilation - a Critical Study of Assimilation Rates and Porometer Rates of Cherry Laurel', *Proceedings of the Royal Society*, B, 102 (1928), pp. 488–533.

Mast, A. R., and E. Conti, 'Commentary. The Primrose Path to Heterostyly', *New Phytologist*, 171 (2006), pp. 439–42.

Matthaei, G. L. C., 'IV. Experimental Researches on Vegetable Assimilation and Respiration on the Effects of Temperature on Carbon-Dioxide Assimilation' in *Philosophical Transactions of the Royal Society*, B, 197 (1904), pp. 47–105.

Mayr, E., *The Growth of Biological Thought* (Harvard, MA: Belknap Press, 1982).

Millett, J., R. I. Jones and S. Waldron, 'The Contribution of Insect Prey to the Total Nitrogen Content of Sundews (*Drosera* spp.) Determined in situ by Stable Isotope Analysis', *New Phytologist*, 158 (2003), pp. 527–34.

Moore, J. R., 'Charles Darwin Lies in Westminster Abbey', *Biological Journal of the Linnean Society*, 17 (1982), pp. 97–113

Morgan, N., 'The Development of Biochemistry in England through Botany and the Brewing Industry (1870–1890)', *History and Philosophy of Sciences*, 2 (1980), pp. 141–66.

Morris, S., L. Wilson and D. Kohn, *Charles Darwin at Down House* (London: English Heritage, 1998).

Morton, A. G., *History of Botanical Science* (London: Academic Press, 1981).

O'Brian, P., *Joseph Banks: A Life* (London: Harvill Press, 1987).

Oliver, F. W., *Makers of British Botany* (Cambridge: Cambridge University Press, 1913).

Opik H., and S. Rolfe, *The Physiology of Flowering Plants*, 4th edn (Cambridge: Cambridge University Press, 2005).

Ornduff, R., 'Darwin's Botany', *Taxon*, 33 (1984), pp. 39–47.

Paley-Marshall, M., *What I Remember* (Cambridge: Cambridge University Press, 1947).

Pfeffer, W. F. P., *The Physiology of Plants, Vol. I.* (1897) trans. A. J. Ewart (Oxford: Clarendon Press, 1899)

—, *Pflanzenphysiologie. Ein Handbuch des Stoffwechsels und Kraftwechsels in der Pflanze* (Leipzig: Engelmann, 1881).

Pickstone, J. V., 'Locating Dutrochet', *British Journal of the History of Science*, 11 (1978), pp. 49–64.

—, 'Discovering the Movement of Life: Osmosis and Microstructure in 1826', *Aspects of the History of Microcirculation*, 14 (1994), pp. 77–82.

Porter, D. M., 'Charles Darwin's Notes on Plants of the *Beagle* Voyage', *Taxon*, 31 (1982), pp. 503–6.

Priestley, J., *Experiments and Observations of Different Kinds of Air. Vol. I* (London: J. Johnson, 1774).

Punnet, R. C., 'Early Days of Genetics', *Heredity*, 4 (1950), pp. 1–10,

Raverat, G., *Period Piece: A Cambridge Childhood* (London: Faber and Faber, 1954).

Reed, H. S., 'Jan Ingenhousz Plant Physiologist. With a History of the Discovery of Photosynthesis', *Chronica Botanica*, 11 (1949), pp. 285–396.

Reynolds Green, J., *A History of Botany in the United Kingdom from the Earliest Times to the End of the 19th Century* (London: J. M. Dent, 1914).

Richmond, M. L., 'Women in the Early History of Genetics. William Bateson and the Newnham College Mendelians, 1900–1910', *Isis*, 92 (2001), pp. 55–90.

—, 'The "Domestication" of Heredity: The Familial Organisation of Geneticists at Cambridge University, 1895–1910', *Journal of the History of Biology*, 39 (2006), pp. 565–605.

—, 'Muriel Wheldale Onslow and Early Biochemical Genetics', *Journal of the History of Biology*, 40 (2007), pp. 399–426.

Romano, T., *Making Medicine Scientific: John Burdon Sanderson and the Culture of Victorian Science* (Baltimore, MD: Johns Hopkins University Press, 2002).

Rothenstein, W., *Men and Memories: Recollections of William Rothenstein 1900–1922* (London: Faber and Faber, 1932).

Rousseau, J.-J., *Letters on the Elements of Botany Addressed to a Lady (1771–3)*, trans. T. Martyn (London: B White, 1785).

Rudolph, E. D., 'History of the Botanical Teaching Laboratory in the United States' in *American Journal of Botany*, 83 (1996), pp. 661–71.

Russell, E. J., *A History of Agricultural Science in Great Britain 1620–1954* (London: George Allen and Unwin, 1966).

Sachs, A., *The Humbolt Current* (Oxford: Oxford University Press, 2007).

Sachs, J. von, *Handbuch der Experimental Physiologie der Pflanzen*, vol. 4 of W. Hofmeister (ed.), *Handbuch der physiologischen Botanik* (Leipzig: Engelmann, Leipzig, 1865).

—, *Lehrbuch der Botanik*, trans. A. W Bennett and W. T. Thiselton-Dyer (Oxford Clarendon Press, [1868] 1873).

—, *Vorlesungen uber Pflanzenphysiologie*, trans. H. M. Ward (Oxford: Clarendon Press, [1882] 1887).

—, *History of Botany (1530–1860)*, trans. from the German edition by H. E Garnsey and I. B Balfour with an original preface (Oxford: Clarendon Press, [1875] 1906).

Salisbury, F. B., and C. W. Ross, *Plant Physiology*, 4th edn (Belmont CA: Wadsworth, 1992).

Sanderson, J. B., 'The Excitability of Plants I', *Nature*, 26 (1882), pp. 353–6.

—, 'The Excitability of Plants II', *Nature*, 26 (1882), pp. 482–6.

—, 'On the Electromotive Properties of the Leaf of Dioneae in the Excited and Unexcited States', *Philosophical Transactions of the Royal Society*, 173 (1888), pp. 1–55.

Sanderson, J. B., and F. J. M. Page, 'On the Mechanical Effects and on the Electrical Disturbance Consequent on Excitation of the Leaf of *Dionaea muscipula*', *Proceedings of the Royal Society*, 25 (1876), pp. 411–34.

Schiller, J., and T. Schiller, *Henri Dutrochet (1776–1847): le Materialisme Mécaniste at la Physiologie Générale* (Paris: Albert Blanchard, 1975).

Schleiden, M. J., 'Beiträge zur Phytogenesis', *Archiv für Anatomie, Physiologie, und Wissenschaftiche Medicin*, 13 (1838), pp. 137–76.

Schwann, T., *Mikroskopische Untersuchungen über die Übereinstimmung in der Struktur und dem Wachstum der Tiere und Pflanzen* (Berlin: Sander'schen Buchhandlung, 1839).

Scott, D. H., 'German Reminiscences of the Early "Eighties"', *New Phytologist*, 24 (1925), pp. 9–16

Secord, J. A., 'Artisan Botany' in N. Jardine, J. A. Secord and E. C. Spray, *Cultures of Natural History* (Cambridge: Cambridge University Press, 1996), pp. 378–93.

Sénebier, J. *Physiologie Végétale* (Geneva: J. J. Paschoud, 1782).

Seward, A., *Memoirs of the Life of Dr. Darwin* (Philadelphia, PA: William Poyntell, 1804).

Seward, A. C., *Darwin and Modern Science: Essays in Commemoration of the Centenary of the Birth of Charles Darwin* (Cambridge: Cambridge University Press, 1909).

Seward, A. C., and F. F. Blackman, 'Obituary Notice. Francis Darwin (1848–1925)', *Proceedings of the Royal Society*, B, 110 (1932), pp. i–xxi.

Shteir, A. B., *Cultivating Women, Cultivating Science: Flora's Daughters and Botany in England 1760– 1860* (Baltimore, MD: Johns Hopkins University Press, 1996).

Simpkins, D. M., 'Knight, Thomas Andrew', in *Dicionary of Scientific Biography* (New York: Scribners, 1971), pp. 408–10.

Small, J., 'The Student as a Synthesizing Organism', *New Phytologist*, 17 (1918), pp. 189–93.

Smit, P., 'Jan Ingen-Housz (1730–1799): Some New Evidence about his Life and Work', *Janus*, 117 (1980), pp. 125–39.

Smith, J. E., *Introduction to Physiological and Systematical Botany*, 3rd edn (London: Longman, 1813).

Spalding, F., *Gwen Raverat: Friends, Family and Affections* (London: Harvill Press, 2001).

Speight, R., *William Rothenstein: The Portrait of an Artist in His Time* (London: Eyre and Spottiswood, 1962).

Stearn, W. T., *John Lindley, 1799–1865: Gardener-botanist and Pioneer Orchidologist* (Woodbridge:: Antique Collector's Club and Royal Horticultural Society, 1999).

Steward, F. C., 'In Memoriam. Frederick Frost Blackman', *Plant Physiology*, 22 (1947), pp. ii–viii.

Stewart, J. G., *Jane Ellen Harrison* (London: Merlin Press, 1959).

Taiz, L. and E. Zeiger, *Plant Physiology*, 4th edn (Sunderland MA: Sinauer Inc., 2006).

Thimann, K. V., 'Plant Growth Substances, Past, Present and Future', *Annual Review of Plant Physiology*, 14 (1963), pp. 1–19.

—, 'The Auxins', in M. B. Wilkins (ed.), *The Physiology of Plant Growth and Development* (London: McGraw-Hill, 1969), pp. 1–19.

Thiselton-Dyer, W., 'Harry Marshall Ward', *New Phytologist*, 6 (1907), pp. 1–9.

—, 'Plant Biology in the "Seventies"', *Nature*, 115 (1925), pp. 709–12.

Uglow, J., *The Lunar Men: The Friends Who Made the Future 1730–1810* (London: Faber and Faber, 2002).

Veak, T. 'Exploring Darwin's Correspondence; Some Important but Lesser Known Correspondents and Projects', *Archives of Natural History*, 30 (2003), pp. 118–38

Vines, S. H., *Lectures on the Physiology of Plants* (Cambridge: Cambridge University Press, 1886).

—, 'On the Digestive Ferment of *Nepenthes*', *Journal of the Linnean Society*, 15 (1877), pp. 427–31.

—, 'Reminiscences of German Botanical Laboratories in the "Seventies" and "Eighties" of the Last Century', *New Phytologist*, 24 (1925), pp. 1–8.

Walker, D. A., '"And Whose Bright Presence" – an Appreciation of Robert Hill and his Reaction', *Photosynthesis Research*, 73 (2002), pp. 51–4.

Walters, S. M., *The Shaping of Cambridge Botany* (Cambridge: Cambridge University Press, 1981).

Walters, S. M., and E. A. Stowe, *Darwin's Mentor: John Stevens Henslow, 1796–1861* (Cambridge: Cambridge University Press, 2001).

Wareing, P. F., and I. D. J. Phillips, *The Control of Growth and Differentiation in Plants* (Oxford: Pergamon Press, 1970).

Went, F. W., 'Reflections and Speculations', *Annual Review of Plant Physiology*, 25 (1974), pp. 1–26.

Wheldale, M., *The Anthocyanin Pigments of Plants* (Cambridge: Cambridge University Press, 1916).

Wilkins, M. B., *Plantwatching. How Plants Live, Feel and Work* (London: Macmillan, 1988).

Williams, B., *Orchids for Everyone* (London: Treasure Press, 1984).

Woodward, F. I., 'Stomatal Numbers are Sensitive to CO_2 Increases from Pre-industrial Levels', *Nature*, 327 (1987), pp. 617–18.

INDEX